The Sixty Second Story
When Lives are on the Line

Janice Landry

Pottersfield Press, Lawrencetown Beach, Nova Scotia, Canada

Library and Archives Canada Cataloguing in Publication

Landry, Janice, 1965-, author
The sixty second story : when lives are on the line / Janice Landry.
ISBN 978-1-897426-51-7 (pbk.)
1. Fires--Nova Scotia--Halifax--History.   2. Fire fighters--Nova Scotia--Halifax--History.
3. Fire extinction--Nova Scotia--Halifax--History.   4. Halifax (N.S.)--History.   I. Title.
TH9507.H28L35 2013   363.3709716'225   C2013-903190-1

Cover design by Gail LeBlanc

Cover image: Halifax Police Department Identification Section and Halifax Regional Fire & Emergency

Photo of author: Paul Darrow

We acknowledge the financial support of the Government of Canada through the Canada Book Fund for our publish---ing activities. We acknowledge the support of the Canada Council for the Arts, which last year invested $157 million to bring the arts to Canadians throughout the country. Nous remercions le Conseil des arts du Canada de son soutien. L'an dernier, le Conseil a investi 157 millions de dollars pour mettre de l'art dans la vie des Canadiennes et des Canadiens de tout le pays. We also thank the Province of Nova Scotia for its support through the Department of Communities, Culture and Heritage.

Pottersfield Press
83 Leslie Road
East Lawrencetown, Nova Scotia, Canada, B2Z 1P8
Website: www.PottersfieldPress.com
To order, phone 1-800-NIMBUS9 (1-800-646-2879) www.nimbus.ns.ca

 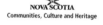 

Dedication

My late father, Basil "Baz" Landry, M.B., was a Halifax, Nova Scotia, firefighting veteran and my biggest supporter. Without his unwavering guidance, I would not have had the fortitude to tackle this large and important journalistic project.

It is with deep respect and admiration that I dedicate this book to all first responders and their loved ones.

The Sixty Second Story is also dedicated to my immediate family: my mother, Theresa Landry; my husband, Rob Dauphinee; his mother, Betty Dauphinee; and, the light of our lives, daughter, Laura Dauphinee.

Lastly, this story is for my father. His legacy lives on, through me, Laura, and this account.

How do you define success?

I think it is summed up perfectly in a favourite quotation of mine, which has become a personal mantra. These eloquent lines truly capture the essence of success and what first responders, both professionals and volunteers, do; th0eir selfless actions literally mean a life is saved, changed, or forever altered.

> *To laugh often and much; to win the respect of intelligent people and the affection of children; to earn the appreciation of honest critics and endure the betrayal of false friends; to appreciate beauty; to find the best in others; to leave the world a bit better, whether by a healthy child, a garden patch or a redeemed social condition; to know even one life has breathed easier because you have lived. This is to have succeeded.*
>
> *– Derived from a 1904 poem by Bessie A. Stanley*

Why are young men told to look in ancient history for examples of heroism when their own countrymen furnish such lessons?

– Sir William Napier (1785-1860)

Contents

Foreword

I vividly remember my father, Basil "Baz" Landry, coming home from a 24-hour-shift on the Halifax, Nova Scotia, fire department with a small paper bag stuffed with delicious penny candy. Dad would stride across our worn linoleum kitchen floor wearing his uniform and a big grin, arms outstretched for a welcome home hug from his only child. He would then hand the treats over, which I would eagerly wolf down.

It was a tradition I deeply relished and it signalled that my daddy was back, safe and sound, from fighting fires. Funny thing, when I was a child, growing up with a parent who was a first responder, I never worried about him not coming home. The reality is, he almost did not make it back to us several times, but I did not know that then. Some of you have friends and/or family members who have worked in this role, yet they have not been that fortunate. My heart goes out to you.

I lost my father in 2006, after he battled heart disease for a year. He had congestive heart failure, a quadruple bypass, open-heart surgery, and struggled, in the end, for every breath. I was there, in a Halifax hospital, at his bedside, when he took his last. He was seventy-three.

His retired colleague Gerry Condon says, "He was one of the nicest, softest-spoken guys. You could really look to him for knowledge and trust. He had the respect of the guys. They would do anything for him, including me. You just know that guy is everything you believe he is." However, Baz Landry's story is one of humility.

As of August 2013, my father is the only Halifax firefighter, in the long history of the city's fire service, to be awarded the Medal of Bravery by the Canadian government. That is why you see the letters "M.B." after his name in the book's dedication. He never used the initials himself.

I knew a lot about this honour, but was unaware of his "first" until I started researching. Our family hopes more Halifax firefighters join him in that rank and soon. It is difficult to express what Dad's accomplishment means to me and our family; this book is my attempt.

I was thirteen years old when it happened. Basil's granddaughter, Laura, is the same age in 2013, when this book was published. It seems an appropriate circle-of-life moment that I am choosing to write about it now.

I did not know until 2012 how close his 1978 rescue came to ending in tragedy. First-hand, eyewitness accounts reveal if one more minute had elapsed at the fire scene, a mere sixty seconds longer, I would never have received another bag of penny candy from the man I adored. That is how close he and the boy he rescued came to dying.

This book will outline that fateful day in 1978, but it begins long before that time, with another far more tragic one: December 6, 1917. The date is forever ingrained in our collective Nova Scotia psyche; it is the moment of the Halifax Explosion. The deadly blast took the lives of at least 1,963 men, women and children, including nine Halifax firefighters, in what remains the single biggest loss of firefighting life, as the result of one incident, in Canadian firefighting history. You will learn about what happened from the first responders' perspectives and also read accounts from two relatives of the nine whom I have tracked down and interviewed; they are also veteran firefighters themselves. The two men discuss what they know, first-hand as family members, about their famous relatives.

At the time of the explosion, and later, during my father's tenure in the Halifax Fire Department, from the 1950s to 1980s, little to nothing was known about critical incident stress (CIS) or post-traumatic stress disorder (PTSD). First responders see, hear, smell and touch what most of us never do, and all of us were never meant to, no matter how much training we have undergone. A portion of this work will review how far this area has evolved and what help is available, whether you are a first responder, a veteran, family member or a loved one. Frank discussions with a number of retirees highlight the deep toll this work can, and does, take.

Sadly, without getting the help they needed, many people working and serving in various front-line careers have turned to alcohol, drugs, gambling or other self-destructive behaviour for escape from the horrors they have witnessed and the stress associated with a job where you are never certain of what to expect when you walk out the front door. Certainly not all people will turn to this type of behaviour, yet others may also suffer from varying forms of depression or other mental health issues.

Thankfully, my father did not follow that path; he avidly golfed, played cards, went fishing, loved the ocean and boating, but, significantly, he rarely discussed his vocation with either my mother or me. I now have a better understanding why. I never thought Dad was affected by CIS or PTSD. I have now come to the conclusion that he and all first responders cannot possibly be involved in a lifetime of dangerous incidents without some kind of personal, emotional or psychological impact.

My father was very humble and private and did not talk about his work or this rescue frequently. He did tell me with great enjoyment lots of stories over the years about the good times – the jokes the firefighters played on one another, the meals they cooked and ate together, the bowling, fishing, card playing, drinks and golfing they enjoyed as a group. It was a close-knit bunch and my father was extremely proud to be part of it. On occasion, he and I would talk about the tough stuff: the calls that were too close for comfort, the lives and property lost, the sights, sounds and smells.

My father did describe to me the details of two fires where he almost did not make it out. The rescue at 3390 Federal Avenue is one of them. One of the few times I saw my father get choked up and emotional about firefighting was when he relived a terrifying scene about his almost running out of oxygen in an apartment building fire just seconds before a fellow firefighter found him, shared his air-pack and escorted Dad to safety.

My father retired in 1988 as a captain after thirty-one years service as a firefighter. After his retirement, Dad and I did discuss certain calls. He knew I was well acquainted with such matters from my work as a television journalist reporting from countless fire and crime scenes. Dad would occasionally tear up discussing if a child had been injured or had died, or when he opened up and described to me the few times he almost died.

He never openly cried, at least in front of me, except at his older brother's funeral. Lawrence Landry was also a first responder, a long-time Halifax Dockyard firefighter and a war veteran. My dad revered him. I now wonder if my father's rather stoic nature about work was actually part of a bigger coping mechanism.

I am grateful he trusted me with some very personal and graphic information and felt comfortable off-loading even a little of it. It will not be revealed here or anywhere. However, I can and will share his description of exactly what it was like inside a pitch-black bedroom the day he saved a baby boy. He did not discuss it with me twice.

Landry is one of two people who can recount the story from that perspective. He is the only firefighter who initially made it into the upstairs room because of the thick smoke and intense heat. A second firefighter witnessed what my father did, in part, from inside. Rob Brown's interview was the last I did for this book. Brown is the only eyewitness from inside the burning home.

The boy in question is now a grown man in his thirties. It had always been a dream of mine to track him down and meet him. It took years of research, phone calls, e-mails, social and traditional media use, and pure luck, but, in the end, I have done just that – the most important interview of this journalist's lifetime.

The Sixty Second Story comes full circle with my face-to-face meeting with him, his mother and grandmother. They describe to me, in detail, the terrifying events that forever changed their family and mine.

While reading this account, please keep in mind my profound respect for all first responders, not solely the special one I grew up admiring and respecting.

In the final few months before his 2006 death, my father handed me a large package of articles, photographs and reports he had collected and kept since 1978. These have been essential in my research; they were the beginning, and are the foundation of *The Sixty Second Story*. My father prophetically said, upon his physically handing the documents over to me, "You will know what to do with these."

I did not give it much, if any, thought at the time because we were in the midst of dealing with his illness and were preoccupied by his many visits to the QEII, which is where he died, on May 2, 2006. His statement about knowing what to do with his collection took me a few years to digest. I was in deep grief after his passing and was not emotionally or mentally prepared to begin the journey that has become this account.

It is now clear the impetus for my work was, in fact, in the very package Dad handed me in that moment. I will never know what he really meant by what he said, but I do believe he was assured his journalist daughter would not only keep his documents safe, but also decide to do the proper thing and tell the whole story of *everyone* involved.

Landry went on to receive high honours, locally, regionally and nationally for the feat which almost cost him his life. The whole story will be revealed here, in detail, from many perspectives, over several chapters. It includes eyewitness accounts from firefighters who attended at the scene and from the family involved. The tale is also told with excerpts from my own personal and private discussions with my father.

Thirty-five-years later, as of 2013, this story will open what remains an emotional wound for some, a very close call that was traumatizing and terrifying. It is a factual, highly researched account told from a variety of viewpoints.

It is my aim and intention that it underlines, for all first responders, sixty seconds definitely matter.

Chapter 1

The Final Alarm – The Fallen Nine

Self-sacrifice is the real miracle out of which all the reported miracles grow. – Ralph Waldo Emerson

Halifax, Nova Scotia, is home to the oldest fire-fighting service in Canada. The first firefighters on record in Halifax were part of The Union Fire Club dating back to 1754, a mere five years after the founding of the port city in 1749. In 1768, The Union Engine Company, the first official fire department in Halifax, was formed, comprised entirely of volunteers. It continued with that name until 1894, when some members were paid part-time and the name changed to the Halifax Fire Department, the precursor of the current Halifax Regional Fire & Emergency.

John Connolly had the honour of being the first chief. The city's growth eventually necessitated the evolution of the department from part-time pay and staff to full-time, fully paid firefighters in 1918. In 1917 there were 122 men in the department – thirty-six were permanent full-time staff with eighty-six partially paid part-timers.

The 1917 Halifax Explosion is a piece of Nova Scotia, Maritime and Canadian history we really wish wasn't; we wish it had never happened. It is the focus of reams of research and countless historical and modern written accounts. Ninety-five years later, as of 2012, our children study it in school. For example, my own daughter made it the subject of her final speech assignment for grade seven.

Despite all that has been read and discussed, you may never have noticed before that it has an unusual timeline and location: 9:06 a.m., December 6, Pier 6. On that day, Halifax became a casualty of World War One. A port with a long military and naval history, Halifax was an important staging ground for ships and convoys in the 1914-1918 conflict. A French freighter, *Mont Blanc*, was gliding through the harbour, in an area locals call The Narrows, which is located in the waterway towards Bedford Basin, near the Macdonald Bridge, the older of two bridges spanning the harbour and shipping lanes. It had previously been in the port of New York and was loaded down with munitions destined for the war effort in Europe. The highly dangerous cargo included 225 tons of TNT, 61 tons of gun cotton, and 2,300 tons of picric acid. The contents were even splayed onto the decks, which held numerous 45-gallon drums of benzol. It was a disaster waiting to happen.

And it did happen, when the Belgian relief ship *Imo*, which had started to leave Bedford Basin, was passing into the same narrowing channel. She was cut off, in a sense, by another vessel making its way into the nearby Dockyard, forcing the *Imo*'s captain and crew to steer east towards the Dartmouth shoreline. In doing so, the two ships collided, with *Imo* ripping into the forward, starboard side of the *Mont Blanc*. The impact tore apart *Mont Blanc*'s foredeck down to the engine room and started the fire, which led to the explosion. It was 8:45 a.m. when the collision occurred and the fire started.

Dark smoke was now billowing in clouds from *Mont Blanc*. Curious crowds of onlookers had started to form along the waterfront, oblivious of any danger.

Captain Lemedec, who was at the helm of *Mont Blanc*, initially signalled for his crew to fight the fire. It quickly became apparent that would be impossible. When Lemedec could not anchor or scuttle the ship because of an equipment malfunction, he ordered his men to the life rafts. They paddled their guts out for shore on the

Dartmouth side. That left the burning vessel unmanned and drifting towards the Halifax shoreline. It was a floating time bomb.

Someone, their identity still unknown, first called for help by pulling Fire Alarm Box #83 at Pier 6, now Pier 8. At this point, *Mont Blanc*, now fully ablaze in bluish flame, had come alongside the wooden pier, also setting it on fire.

Historian Don Snider, a retired administrative captain with the Halifax Fire Department and a collector of fire service artifacts and memorabilia, clarifies the exact location of Alarm Box 83 as "the corner of Roome Street and Campbell Road, which is now Barrington Street. The box would have to be opened and pulled."

It was now shortly before nine a.m.

Then Constant Upham, who owned a general store on Campbell Road, telephoned the fire department. Historical accounts describe him calling around to various stations.

The nine firefighters who initially responded to Pier 6 were not expecting anything out of the ordinary that morning because, as Snider explains, fires were fairly typical on the waterfront in 1917. Coal was loaded onto ships along the docks and the dust and debris often caught fire.

One person who did sense impending disaster was Vince Coleman, the now legendary telegraph dispatcher who worked at the Richmond Railway Station in Halifax's North End. Coleman famously alerted the dispatcher in Truro with his brief report: "Hold up the train. Ammunitions ship afire in harbour making for Pier 6 and will explode. Guess this will be my last message. Good bye, boys."

First responders raced to the waterfront from varying locations around Halifax. Later, they came from across the province and beyond. Popular historical accounts underline where they came from, how they travelled and who was involved, from the outset.

At the West Street fire station, Billy Wells ran outside and jumped into *The Patricia*, an American LaFrance motorized fire pumper, the first of its kind in Canada: the most up-to-date rescue vehicle money could buy and the only motorized truck in the department.

Wells, by all reports, was anxious to respond to this first call of the day in 1917. His brother Claude usually drove 60-year-old fire chief Edward Condon in the senior officer's new McLaughlin Buick roadster. On this day, Claude was fortuitous to have the day off. Filling in as the chief's driver was 41-year-old deputy chief William Brunt.

At West Street, the fire crew who boarded *The Patricia* was comprised of 32-year-old captain William Broderick, 25-year-old hoseman Walter Hennessey, 21-year-old hoseman Frank Killeen, 35-year-old hoseman Frank Leahy, and captain Michael Maltus, whose age is unknown. Maltus was filling in for another unnamed crew member who took the day off because he was ill.

Thirty-four-year-old hoseman John Duggan was working from the Isleville fire station, located in the north end of Halifax on Gottingen Street. Duggan had to take the time to harness his horses to the #4 hose wagon. Horse-drawn wagons and carts were the standard mode of transportation for firefighters back in 1917. After the wagon was quickly harnessed and loaded down with hose, Duggan started towards the fire scene at full gallop.

When Captain Broderick and the crew of *The Patricia* arrived at Pier 6, *Mont Blanc* had drifted alongside the pier and both were on fire. Reports indicate that, by then, the fire was so powerful, and the heat so intense, firefighters were forced to turn their faces away to shield themselves.

The hose wagon, manned by Duggan, arrived at approximately the same time as the two senior officers, Condon and Brunt, in the roadster. Captain Broderick gave his initial report to the chief, summing up what he had witnessed since his arrival. Chief Condon immediately ordered hose line to be run down to Pier 6 and had Fire Alarm Box #83 pulled a second time. The situation was deteriorating and it was obvious to the chief they needed help.

As a result of that second alarm, at the Brunswick Street station, 65-year-old hoseman John Spruin, a "call fireman" who lived next door to the station, got dressed in his gear and responded even though he was retired.

According to reports, Spruin had been a respected firefighter with a lengthy and distinguished career. Since his retirement, he had been working as a janitor at City Hall. As a call firefighter, he was allowed to respond to second alarms. He was also on a hose wagon as it sped up Brunswick Street towards Pier 6. Like his peers who had already responded, Spruin had no idea what awaited him, just that it was bad enough that a second alarm had been sounded.

At the fire scene, firefighters had taken lines from both the hose wagon and *The Patricia* and ran them across to the pier. Wells was positioning the engine up to a hydrant when a first, smaller explosion took place. Accounts describe it as "a muffled roar and the ground shook, knocking everybody off their feet."

The following excerpt from a historical account, published in the *Atlantic Firefighter* in December 1917, describes the second blast that we have come to know as the Halifax Explosion:

"Seconds later the *Mont Blanc* disintegrated in a blinding white flash. The explosion was heard and felt as far away as North Cape Breton, 225 miles from Halifax. It was 9:06 a.m. Billy Wells was ripped from his seat still holding half of the steel and hardwood steering wheel in his hands. Chief Condon's pride and joy, his shiny black McLaughlin Buick, was somersaulted backward and lay upside down completely wrecked, both Chiefs dead. The crew of *The Patricia*, except Billy Wells and Frank Leahy, were dead. *The Patricia* was wrecked and destroyed but still on her wheels. John Spruin was hit on the head with a piece of flying debris and knocked off the hose wagon on Brunswick St.; he struck his head on the curb killing him instantly.

"Billy Wells found himself sitting up some distance from *The Patricia* with the steering wheel still in his hands. All he had on were his pants and one boot and the muscles were turned [torn] from his right arm, but he was quite conscious. He could see *The Patricia*. Next a tidal wave carried Billy up the side of Richmond Hill, knocking him out. When the wave came back down the hill, Billy got tangled up in the telephone wires and wreckage and almost drowned, but came to, and sat there confused. Next came a 'black rain' of unconsumed carbon from the explosives. It fell for about 10 minutes; an oily soot that coated everything in the area of destruction: rubble, bodies, faces, clothing, everything was coated with this back finish."

Covered in soot and blood, Wells eventually started walking. In bad condition, he briefly collapsed but regained consciousness and started walking again. All the while, he was still gripping half of the steering wheel in his hands.

Don Snider confirms Halifax Regional Fire & Emergency still owns half of the steering wheel, the portion attached to *The Patricia*. It's not known what happened to the piece Wells clutched during his wanderings.

Wells continued walking and was headed south along Campbell Road (Barrington Street). In fact, he walked right past *The Patricia* but reportedly did not see any of her crew. He continued past the #4 hose wagon, which was destroyed. The horse was found dead nearby. There was no sign of Duggan anywhere. His remains have never been found. The collar of his horse was found near Oxford and Almon Streets, slightly more than three kilometres away.

As motorized fire engines were only just starting to become available, horses were the most commonly used method of moving men, hose and equipment to fire scenes. The era of horse-drawn hose wagons ended in 1929. It was during that year the following poem appeared in a local newspaper:

> *The Fire Horses Farewell*
>
> *We played the game and played it*
> *Square, at every call of the gong;*
> *We gave our speed, upheld our*
> *Breed, now our life's not worth a song;*
> *Our time is past, the die is cast;*
> *Perhaps it's just as well.*
> *So to our Halifax friends and the*
> *Fire Brigade, we neigh our last farewell.*

After passing the deceased horse and destroyed hose wagon, Wells continued on to Condon's roadster, which was lying in a mangled heap. He saw two children who were alone and frightened. The wounded firefighter took care of them until he came upon two sailors from the warship *Niobe*. Wells asked the sailors to help the youngsters. With the kids taken care of, Wells sat down, weak and injured, to rest. He was having trouble going further. At that crucial point, a salvage wagon crew came by and rescued him.

They immediately searched the entire area for any surviving firefighters. The only one they found alive was hoseman Frank Leahy, unconscious and severely injured. Both he and Wells were taken to hospital for treatment. Only one of them would make it. Leahy died twenty-five days later, on New Year's Eve, 1917, as a result of complications from his extensive injuries. He became the final member of The Fallen Nine.

His obituary reads: "Frank Leahy, a member of the H.F.D. and last of *The Patricia* crew, died of injuries Dec. 31st, passed away this evening at the Victoria General, where he was taken on the day of the great explosion, grievously wounded. Mr. Leahy died, as did his chief and comrades, at the honourable post of duty, having, on the fatal morning, responded to the call of fire. He had for four years been a member of the H.F.D. and was well liked by his fellow members, full of courage, of generous impulse, upright and kindly. It was recognized from the first that his injuries were very serious indeed, but medical skill did its utmost to save him, his own patience and strength of will assisting the fight. However, all was unavailing and he passed away on the evening of the last Sunday of the year, to a world wherein faithfulness to duty is entailingly rewarded."

Billy Wells was the sole survivor of the ten firefighters who initially responded to Pier 6. He spent five months in hospital recovering and recuperating. He worked as a firefighter until 1926. After his retirement he worked for Oland Breweries until 1948. Wells died in 1971.

One of the last interviews with Wells appeared in *The Mail Star,* December 6, 1967, on the fiftieth anniversary of the Halifax Explosion, four years before Wells's passing. He was eighty-seven and lived on Agricola Street, not too far from where the explosion occurred. In that article, Wells confirmed the fire crew went racing to Pier 6 unaware of the "powder keg" that awaited them: "We didn't know the ship was carrying munitions." He described the horrendous morning: "It was about twenty minutes to nine when we received a telephone call at the West St. fire station saying there was a ship on fire at Pier 8 [then 6]. Our fire engine, *The Patricia*, had a crew of 8 men. I was the driver and we immediately rushed down to the Pier. The ship was almost alongside the dock and the multi-coloured flames shooting from her decks to the sky presented a beautiful sight."

The firefighters thought the ship's crew was still on board, according to Wells. They started to unroll the hose: "That's when it happened ... the first thing I remember after the explosion was standing quite a distance from the fire engine. The force of the explosion had blown off all of my clothes as well as the muscles from my right arm."

Wells said he was conscious when the ensuing tidal wave swept over him: "After the wave had receded, I didn't see anything of the other firemen, so I made my way to the old magazine on Campbell Road ... the sight was awful with people hanging out of windows dead. Some with their heads off and some thrown onto the overhead telegraph wires. I was taken to Camp Hill Hospital and lay on the floor for two days waiting for a bed. The doctors and nurses certainly gave me great service."

Despite the trauma over losing so many comrades, all fire companies, stations, off-duty and call firefighters turned out, en masse, to help in the rescue effort. They also fought dozens of out-of-control fires burning throughout the city.

Most homes and businesses in 1917 were old or wooden or both, which meant the fires spread rapidly. Many people were trapped in the collapsing and burning buildings and died in the fires or were killed from flying or falling glass and debris. A second excerpt from the *Atlantic Firefighter* describes the immediate and valiant rescue efforts:

"The Halifax Fire Department worked through the day under the worst conditions, wet, cold, without food, without assistance. The 32 career firemen and 120 volunteers who survived the explosion pushed themselves and their apparatus to the limit. The only motorized unit, *The Patricia*, was wrecked and out-of-action. However, horse-drawn units were still available, five active and two reserve steam engines, two chemical engines, eight hose wagons and four ladder trucks. Just in front of the west gate of the Wellington Barracks [now CFB Stadacona] on the corner of Gottingen and Macara Streets, a grocer's shop caught fire and burned to the ground. All about were flimsy, wooden houses, which would have burned like tinder. It was at this point that the firemen would have rendered their most efficient service in checking the spread of the flames. Their hard work, at this point, undoubtedly kept the fire from spreading south and averted a great danger."

As news of the explosion spread, firefighting crews from across Nova Scotia raced to help. Relief efforts and trains from across Canada and the Eastern United States were also on their way. A train from New Glasgow with a steamer, three thousand feet of hose and forty firefighters was dispatched within an hour. It stopped in Stellarton and Truro to pick up more firefighters and supplies, arriving in Halifax later that afternoon. Other trains stocked with men and equipment from Kentville, Amherst and Sydney came that evening.

The relief firefighters scoured the ruins looking for survivors and helped to patrol the streets of the city's north end, giving Halifax firefighters some relief and rest. Historical accounts indicate many of the fires were put out by morning, but others raged on for some time. A snowstorm and bitterly cold temperatures blanketed the city the following day, adding to the desperation and misery.

Historian Don Snider says, while *The Patricia* was blown up in the blast, a second motorized pumper had arrived in the city not long after the explosion, having been ordered a year or two before. The surviving firefighters would have used the new vehicle to battle the ensuing fires in the devastated city. In case you are wondering about the actual power of the American LaFrance vehicles, Snider states *The Patricia* had 67 horsepower and the capacity to pump 750 gallons per minute. Snider confirms it was eventually salvaged, sent away for repairs and placed back into firefighting service in the Chester, Nova Scotia, area. However, over the years, *The Patricia* changed locations and its importance somehow diminished. This crucial part of Halifax's history, sadly, was not preserved. "It is criminal that it no longer exists today," exclaims Snider. "Of all the vehicles! Get rid of anything else but that!"

The Patricia is a priceless yet now lost artifact, which underscores the devastation, loss of life and incredible story of the fire service's crucial role in this sombre chapter of Canadian history.

Chapter 2

Legacy of Lifesaving

The factual and historical accounts of the heroism stemming from the Halifax Explosion live on through direct descendants, who themselves have carried the torch in more modern firefighting times.

For example, hoseman John Duggan has a blood relative who was also a respected Halifax firefighter. Bernie Harvey started in the fire department in 1954, rising through the ranks to eventually become a district chief and platoon chief. At some points in his career, Harvey worked out of Central Headquarters, which was located on West Street. He retired after thirty-five years of service in 1989.

On the day I meet Harvey for our lengthy interview, he welcomes me into his very neat and tidy north-end Halifax home. We sit in his kitchen, the focus of all activity for most Maritimers, where he proceeds to explain how he is directly related to one of The Fallen Nine, his grandfather.

"My mother, Mae (Duggan) Harvey, that was her father, John Duggan. John worked out of the Isleville Street station. He drove the hose tender, which had no pump on it or water. Back in those days, they just carried hose with a wagon and a horse. They would go to the scene and get the hose ready for the pumper. *The Patricia* was the first motorized fire apparatus in Canada. Everybody, including me, thought it was stationed on Isleville Street, but it was on Gottingen Street, in the vicinity of where the former North End Beverage Room used to be. They used to call that area Isleville at the time."

For hoseman Duggan, it was his final call from Isleville.

His grandson, Harvey, recounts the fateful day. "The call comes in. When John Duggan was alive, there was no radio responses in any apparatus. I remember when I started there were no radios. You had to go right to the scene [without knowing what the situation was]. But, before that, Duggan had to get the harness on the horse, get the horse tender hooked up and proceed to the location where the call came in from … He was responding to the fire call when the explosion happened. There is a conflicting report that he was laying feed lines when the blast occurred. I know they never found his body, mainly because, after the explosion, a tidal wave was created in the harbour. The wave came in and washed his body out to sea and they never recovered it."

Harvey says he also worked around ammunition ships, like *Mont Blanc*. From about 1965 through to the early 1970s, Harvey used to perform fire watch on the Halifax waterfront. It would require him to be on duty all night. "I worked in Fire Prevention down at Pier 28, by Point Pleasant Park. It would be me and one or two other firefighters. In the event of a fire, we would call the fire department and pull the box in the shed.

Harvey recounts one terrifying evening while he was on watch down at the pier. "We were there and an ammunition ship is coming in. The stevedores were there and they [the ship's crew] used to throw the lines over to the men on the pier; down at Pier 28, those piers are now all concrete. When the ship came in, it hit the concrete wall of the pier and the sparks flew all over the deck of the ship. Its side was all scraped. You want to have seen those stevedores running. You could see the heels of their shoes hitting the backs of their heads running up the pier. They were scared to death. We stayed there; nothing happened. Thank God. The ammunition was stored below. I can still see them running."

With that revelation, Harvey rises and leaves the kitchen to go to a nearby room and bring back a family treasure and heirloom. It is a framed certificate formally presented to him by a local firefighting union to help mark and honour the valour of his late grandfather. It reads: "Respectfully presented to the Duggan family, on December 6, 2004, on behalf of the members of Halifax Professional Firefighters, in recognition of the ultimate sacrifice made by hoseman John Duggan, who died in the line of duty, on December 6, 1917, while performing his duties as a firefighter."

Harvey has proudly hung the framed certificate in a place of honour in his living room, a visual and daily reminder of the lasting impact of hoseman Duggan's sacrifice.

Another descendent of one of the nine fallen firefighters, who also speaks with great pride about his family's history, is Gerry "Crash" Condon. Condon is directly related to the superior officer from 1917, chief Edward Condon.

Sixty-seven-year-old Gerry Condon is the late chief's great-grandson. He was twenty-two when he started as a Halifax firefighter in 1969, when the city's first amalgamation occurred. "When I started in the fire department, I would work part-time at Sears. As I finished my shifts, I'd go over to my grandfather's on Bayers Road. We'd play cards and talk. I loved my grandfather [Frank Condon], I can tell you that. He would discuss with me what it was like to be a firefighter. He was a captain. He was my mentor. He was everything that I wanted to be on the fire department. I wanted to be like my dad too, but I wanted to be like my granddad on the fire department. He was respected."

Condon served thirty-seven years before retiring in 2006. His firefighting lineage was and is impressive. Condon explains his family's history with great relish, while I had tea with him and his wife in their immaculate kitchen: "His [the chief's] name was Edward Condon. His middle name was Francis. My middle name is Francis. My grandfather was Francis Edward. My father was John Francis, and I'm Gerald Francis. That's the way it went."

History does have a way of repeating itself within this fire-fighting family. Condon retired as a captain, like the grandfather he adored. "I'm very proud. My great-grandfather's in the history books in Canada, no, around the world! At the time, it was the biggest man-made explosion ever."

Condon says the chief had three children: Francis (Frank), Gerald and Catherine (Kitty). The Condon family actually lived at a fire station, which was located at the site of the Metro Centre, on Brunswick Street. "When the chief died, he never had a mark on him, according to my grandmother," Condon says. The chief's final resting place is in Holy Cross Cemetery in the city's south end.

Condon picks up the story with the aftermath of the blast. "A couple of days later, in the snowstorm, they had Boston sending stuff and other places, too. The city fathers assigned people to do tasks to help keep the city running. Chief Condon's daughter, Catherine, my great-aunt Kitty, had a husband who was also in the fire department. She had studied at Miss Murphy's Business College and was the fire department's bookkeeper at the time of the explosion."

In her day, Kitty told Gerry many stories about her father, whom she described as "very calm, gentle and relaxed." Condon also says Kitty was very humble about the fact she was asked to play a larger and key role in day-to-day fire operations after her father died. The next chief, John Churchill, was appointed six days later. "She created the shifts, schedules, payrolls, did the books and budget; she basically managed the department. Remember now, the city was in chaos. They had the whole Commons full of tents and injured people. They had people up at the Citadel. They had people dying or freezing to death. This went on for weeks."

After they named Churchill as Condon's successor (he served in that post from 1917 to 1945), Catherine stayed on serving as secretary for the fire department. Condon says she was in her twenties during this era. "And, not ever, ever, ever in her life did she once say, 'Oh, I did all that.'"

Condon is part of a long line of firefighters in his family who have served Halifax They include Kitty's famous father, her husband George Butler, and Chief Condon's two sons, Gerald and Frank (who began as hosemen and went on to become a captain and a lieutenant) and Gerry himself. "That's all the Halifax firemen we know of ... before that, we don't know," Condon concludes.

What we do know is these personal, family accounts, shared by blood relatives and recorded here, are a crucial element in staying connected to The Fallen Nine and to a gut-wrenching chapter in the history of Canada's oldest fire service. Both retired firefighters, coincidentally, also have a connection to my family – they both worked, over many years, with my father.

Chapter 3

Remembering the Fallen

Despite the fact the Halifax Explosion remains, as of 2013, the largest loss of firefighting life in Canadian history, surprisingly, three-quarters of a century actually passed before The Fallen Nine – William Broderick, William Brunt, Edward Condon, John Duggan, Walter Hennessey, Frank Killeen, Frank Leahy, Michael Maltus, and John Spruin – were officially honoured. The commemoration ceremony of Halifax's Fallen Firefighter Monument, complete with flag bearers and the fire departsment honour guard, occurred on December 6, 1992, seventy-five years after the explosion.

The event took place in front of Station #4 on Lady Hammond Road, located in the city's north end. It is not far from Fort Needham Memorial Park, a large, grassy hill and beautiful open space that's the location of the annual Halifax Explosion bell ceremony. It takes place each morning of December 6, at exactly 9:06 a.m., to mark the deadly blast.

On the front of the dark, stone monument is an etching of a turn-of-the-century firefighter wearing ornamental dress, the kind typically seen at parades rather than at fires. Don Snider says the picture does not depict how Halifax firefighters would have been dressed when they responded to the alarm at Pier 6. He explains that most firefighters in Halifax would have worn long oil-skin rain-coats, sou'wester style hats and rubber boots.

A few, Snider adds, would have had leather helmets, but certainly not all of them. The firefighter on the monument is also holding a nozzle. He does not have breathing apparatus on because that did not become common practice until the late 1950s. The gear described was certainly little protection against the conditions firefighters routinely faced. Yet there was nothing typical about what happened to The Fallen Nine, whose names are carved into the monument's face. On its rear are the names of every Halifax line-of-duty firefighting death, beginning with the first one, which dates back to the 1800s.

The complete and official list of all the line-of-duty deaths is compiled by Snider and a team after lengthy research and a stringent checklist has occurred. Their names are found in the Postscript.

The official list started to take shape after Snider searched old city records. "I went to the library and the archives and into old city Fire Ward Minutes from meetings. That's how we started to figure it out. That began years ago, and since then, we have discovered new ones. Their names are out of order on the stone because we place them on it as we discover and approve them. We have to get all the facts first and make sure it's accurate. These are people who died at the fire scene or days later from injuries sustained at the fire," explains Snider.

The stark monument had humble beginnings. It came into being because of the inquisitive son of Dave Singer, a retired Halifax firefighter. Singer says his son Jeffrey was in high school working on a class project in 1989, when he asked his dad some key questions. Snider, who joined the fire department in 1962 and worked for thirty-five years in his role before retiring in 1997, also recalls how Singer's research started. "Dave's son, Jeffrey, asked his father, 'Dad, I'm doing some research on the Halifax Explosion. What did the fire department do?'"

Snider adds, "We had all heard about the survivor and the chief who was killed. But it was all sketchy. Their [Dave and Jeffrey's] conversation kind of went like this: 'Well, son, the fire truck *The*

Patricia was blown up, Billy Wells was the survivor and some guys got killed.' To which the son said, 'Dad, let's find out about it.' The boy and his father basically started doing all the research. They came up with the whole story about the nine being killed, and from there, it just skyrocketed."

Singer, fascinated by the task of discovering more history, kept working and researching long after his son's project was completed. Singer says he spent years on it and, once he discovered its magnitude, concluded, "This just isn't right. These men died and there's not a monument, not a service." That realization spurred both Singer and later Snider on; the men, joined by others, wanted to create a lasting memory in a permanent marker of valour.

Dave Singer approached then Halifax fire chief Tom Power to ask about creating a lasting homage to The Fallen Nine. The idea of the monument was born. Don Snider says it was staff at Heritage Memorials who designed and created the stone masterpiece. The cost of the monument, twenty years ago, was approximately $12,000.

The large sum required a fundraising team. Firefighters, department staff and employees and their loved ones, diligently worked on the memorial project. Singer approached Snider and other individuals to help raise the money. Singer even designed a pin to be sold as a fundraiser.

"It had nine stars in the middle, for the nine who died. There is a wreath, which designates high esteem, and the Maltese Cross represents the fire. There is also some white on the pin, which represents the snow the next day. Blue is also part of the pin, which is in there for the water and the tidal wave that hit. The dates on the medal are 1917-1992, marking the seventy-fifth anniversary of the explosion," says Snider.

The pins were sold to firefighters, staff, families and friends. I actually own one of the twenty-year-old pins. I came across it while researching for this book. It was my father's. He kept it in a little blue leather and velvet-lined box, along with some of his other treasured belongings. It was stashed away in an old briefcase that he gave me before his passing. I never understood the pin's significance until I spoke with Snider and Singer.

Dedicated fire department administrative employees who worked in the office with Snider helped him send out hundreds of letters asking for donations to help to pay for the monument. "We got $1,000 from the Province of Nova Scotia; all kinds of organizations gave us $100, $500 and $1,000. We raised about $25,000. Everything was covered, including the commemoration service and reception that followed," says Snider.

The large gap in time that passed between the loss of firefighters in 1917 and the official commemoration ceremony in 1992 is a bit of a sore point with Snider and others. He is now satisfied that the nine men have finally received the recognition they deserve.

"The sad part about it was everybody knew about the explosion, you just didn't know what happened with certain organizations. The army and navy also had an important role, as did the police. But ours is so impactful because of the death toll. It was phenomenal. You know what I think people have to understand most? They have to realize that a huge number of our members died, yet the rest [of the firefighters] had to go on. It wasn't just during the explosion and its immediate aftermath. What happened afterwards is that all these older houses that were using coal and wood, they all burned up; all these houses ignited. Fires were burning everywhere, and these guys that were left, they had to move forward. Nine of their comrades were killed and they still had a role to keep going," Snider says vehemently.

A stickler for detail, Snider was visiting a firefighter at the Lady Hammond Road station, the site of the monument, when the fellow was painting the kitchen and mentioned to Snider that he wanted to "dress the place up a bit." Snider told his friend he had an idea. He suggested they take down the eight-foot corkboard on the wall, clean up the area and prepare it for a special memorial Snider had in mind. "I said, 'I'm going to start to get you some pictures and we'll do a write-up of all the line-of-duty deaths named on the back of the monument.' I gave them a print of the fire truck, *The Patricia,* and we got pictures of Chief Condon's car, with him in it before the blast, and we have another photo of the car all smashed up afterwards. They are all framed. We also have a listing of all the names and a little story of how all those died in the line of duty."

Snider's idea discussed in the kitchen that day became a Wall of Remembrance, a significant and visceral link to Halifax's storied fire-fighting past. If there was any question whether it was worth all the effort, that was put to rest on the day the firefighters were hanging the last picture on the wall. Snider says something special occurred. Captain Derek Jones, who was on duty, looked outside and saw a couple standing by the monument taking photographs. Jones decided to go outside and talk with them. They told him their father's name was on the back of the monument. Jones invited them inside the station to see the newly finished wall, photos and write-ups. They were floored to see how much care and concern was now being taken in the memory of all the fallen, including their beloved father, whom Snider did not name in recounting this story.

Another Halifax firefighter, 44-year-old Chris Camp, treasurer of the International Association of Firefighters (IAFF) Local 268, says since the monument's official unveiling in 1992, he had also become concerned about the level of attention the fallen were – or, more specifically, were not – receiving.

Camp says he had noticed the dwindling number of people attending the annual firefighter's memorial after municipal amalgamation in 1996, when thirty-eight fire departments became one. "There was a big launch in '92. Four years later, we're into amalgamation and we're [the fire service] a large unit. We'd joined Dartmouth, Sackville and the other areas. Everyone was so concentrated on helping things move forward that a lot of the past was forgotten about. I showed up here one December 6th and there was the mayor, the chief and maybe six other people. It annoyed me that we had let it get to that. It should have never have gotten to that. I said, 'Let's put a committee together. Let's start working on it.' At first there wasn't a lot of response, but two years ago [in 2010] the premier [Darryl Dexter] actually showed up at our event and people were like, 'Well, the premier is here, we must be important.' That was a watershed moment. It's getting back to where it should be."

In order to stay where it should be, the public is invited and encouraged to attend the annual service for the fallen firefighters. Each December 6th, the honour guard leads the procession, which involves a pipe and drum band. "If there is a family member present, we will lay a wreath for that line-of-duty death. We always also make sure we recognize the nine members killed during the explosion. All three levels of government lay wreaths, and after the service, we have a reception for everyone inside the fire station," Camp explains.

The firefighter hopes the numbers attending the memorial stay strong and continue to grow, as more people become aware of the incredible history of firefighting in Halifax. Camp's own interest in the past stems, in part, from his own roots. When Camp started working in 1990, he was stationed with mostly senior firefighters at the Bayers Road station. "I don't think anyone had less than twenty-five years on the job. I was fortunate."

Camp's appreciation of those who have served, and those who have fallen, had its beginnings in his early days – living and working alongside veteran Halifax firefighters, sharing meals, stories, and, responding, side by side, when the alarm sounded. It is an intense bond that does not end, as families of fallen or deceased firefighters have discovered.

It was out of that very station, in 1978, that my father and his first responder colleagues went to a fire call that is the crux of this book.

Chapter 4

The Rescue at 3390

The frantic call for help happened at 18:40, twenty minutes before seven in the evening. The 24-hour clock is the official clock used by many first responders and the military. It appeared this way on the official Daily Report document prepared for the fire department back in 1978.

Copies of official records surrounding the rescue were given to me for my research and use by three sources: initially from my father, who kept many factual and historical documents about the rescue and ensuing media coverage; next, from Don Snider, who has spent countless unpaid hours trying to save key records and data and who knew Landry; and lastly, from the office of the current Halifax Regional Fire & Emergency chief, Doug Trussler, through Darlene Ellis, the coordinator of the chief director's office, who told me she also knew my father. Other essential assistance was provided by the chief's administrative assistant, Linda Dodge.

All my sources have been invaluable and are appreciated. They have aided in the accuracy of this and ensuing chapters. The timeline and details described come from documents and eyewitness accounts stemming from my interviews with firefighters who attended at the scene and the family of the rescued boy.

The urgent need to permanently record these crucial perspectives quickly became apparent once I began my research. The number of firefighters who were actually present in 1978 and who are still alive to tell their stories is sadly dwindling. All the men I have spoken with have long since retired and range in age from their sixties to their late eighties, and all are still able to recount with great clarity and in detail how the evening unfolded.

This is only one story, but it is *the* story of the daring rescue that led to Baz Landry receiving Canada's Medal of Bravery. It was recorded as follows in the official dispatcher's report:

Timeline: 18:40 October 2, 1978. 3390 Federal Avenue, Halifax.

Incident Response Number: 782823

Time of call: 18:40

Nature of call: 5-1/5-2 [Snider explains, back in 1978 a 5-1 meant a tentative working fire and 5-2 indicated a confirmed working fire.]

How received: Phone

Location of call: 3390 Federal Ave

Remarks: Sound of caller's voice: Excited

Caller: Male

Weather conditions: Fair

Response: 7-7A-R2

The response numbers and letters in the previous line reveal there were three fire apparatus and crews attending on October 2, 1978: #7 Engine, #7 Aerial, and Rescue #2, which together formed #7 Company working out of the Bayers Road station. Halifax has since renumbered all its fire stations after amalgamation when it became the Halifax Regional Municipality (HRM) in 1996.

The home in question, 3390 Federal Avenue, is located off the lower portion of Bayers Road, which is a main artery in and out of the city. The neighbourhood is off Romans Avenue, primarily comprised of two-storey, attached, row housing units; therefore, a major concern of the responding men was to keep the fire from spreading into other adjoining and nearby homes.

In 1978, the captain in charge at Bayers Road was Frank Walsh, now deceased. Walsh was affable, fun-loving and a very close friend of Landry's; the two often went deep-sea fishing together in their off-hours. Filling in for Walsh that evening was Gordon "Champ" McIsaac, a lieutenant like Landry, but who was acting captain for that shift. McIsaac, also now deceased, was a close friend of both men. Don Snider says Walsh and he were two of the honorary pall-bearers at Champ McIsaac's funeral on October 19, 2007. About three weeks later Walsh passed away on November 10, 2007.

One of the firefighters who attended in 1978 with Landry and McIsaac is 66-year-old Joe Young, who retired in 2000 as a platoon chief. Like Gerry Condon, Young has a rich firefighting lineage in his family: his father, Harris William Young, was a volunteer chief in Fairview in the early 1950s. His son, Allan, joined Halifax Regional Fire & Emergency in 1998 and as of 2012 was working as a lieutenant at Station #11 in Sackville.

Joe Young was the driver of Rescue #2 and of Lieutenant Landry the night of the fire. He says he drove Landry to and from countless fire scenes, but this is the one that still moves him to tears when he talks about it, more than three decades later, in the dining room of his lakefront home. Young explains how the rescue vehicles responded on October 2. "Number 7 Engine is out first [from Bayers Road] and we're [Rescue 2] out second. Seven Engine went into Romans Avenue and took a right onto Federal Avenue. I also went the same way, left onto Romans, right onto Federal." A third rescue vehicle also raced to the scene: #7 Aerial.

Halifax's Federal Avenue is laid out in a semi-circle and can be accessed by turning right off Romans Avenue at two different points, one closer than the other. On this day, both #7 Engine and Rescue #2 turned down the street at the first of the two access points, so they ended up, according to documents and to Young, facing the same direction and parked near one another on Federal Avenue. The aerial truck was parked a little farther away.

The scene is hand-drawn in a second official report called Halifax Fire Department – Details of Firefighting Operations, given to me by Don Snider during a 2012 meeting at his north-end Halifax home. It was a very emotional moment for me to receive it from Snider because the sketch is actually drawn by my father. I was very used to seeing Dad's writing and drawings because he was a

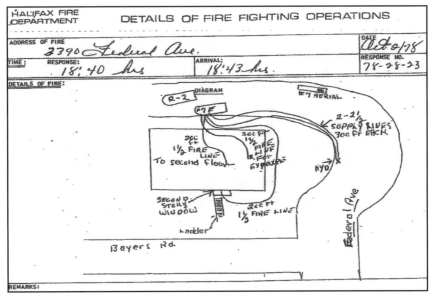

The sketch showing the fire scene at 3390 Federal Avenue. Landry's sketch does not include Romans Avenue, which is the left-hand turn off Bayers Road the first responders took to reach Federal Avenue. (Courtesy of Baz Landry's estate and Halifax Regional Fire & Emergency.)

carpenter in his off-hours and often drew floor plans and other construction projects which were always in various stages of progress. His older brother, Kenny Landry, was a master carpenter and taught him the ropes. Dad also used to draw little funny faces on personal notes for me over the years, many of which I have kept. So when I saw this fire department sketch sitting among a pile of papers and records on Snider's kitchen table, I was immediately overwhelmed; I knew by the way it was drawn that only one person could have done it. It was a rare, personal connection after my father's passing.

Filling in more of the details a one-dimensional drawing simply cannot convey, Young further explains what happened en route to the fire. The firefighters on Rescue #2 were told via radio by Captain McIsaac, who had arrived first on #7 Engine, to expect "smoke condition," meaning there was smoke showing from the home. They were also instructed by McIsaac to "'Run us a feed line.'"

Young says of their initial response plan, "That means the second unit in [Rescue #2] is going to take a hydrant, which is to wrap the hydrant with a hose, drive down and feed the water into #7 Engine, which is a pumper. That was going to be our job. We worked as a company. We knew automatically that if they were running a line we were running a feed. But when we stopped ..." Young trails off, tearing up and having a hard time continuing. At this point, I am unsure why. We are seated in his dining room, overlooking a lake and his beautiful property.

I say, "This is emotional for you."

It is also emotional for me.

Young has trouble regaining his composure. I get up from my seat and walk over and hug him, saying, "Thank you so much." I am grateful for his willingness to share a story that obviously still moves him, dredging up old ghosts and stories from his firefighting past. I am impressed by his willingness to take down walls for me, the daughter of a man, I was about to learn, Young had revered.

"Baz saw something," Young says.

"What do you mean?" I ask.

"I don't know. I can't tell you," he answers.

I come to understand that he believes, wholeheartedly, that Landry saw something when Young parked Rescue #2 next to a fire hydrant on Federal Avenue to hook up the lines to feed water to the pumper. During those brief moments, Young believes Landry sized up the situation and knew what he was going to do.

"We could see the back of the building where the fire was and where the smoke was coming out. Basil said to me, 'When you get up there, Joe, get a ladder in, around back.' As soon as we stop, Basil gets out of the pumper. I jumped out. The ladders on the pumpers were fairly high up, and I was in pretty good shape back then. We grabbed the ladder."

The "we" Young refers to includes Frank Zwicker, another firefighter who was present that day.

Before Zwicker and Young manoeuvre the ladder around the back of the burning building, Young first completed his main assigned task.

"I can remember it to this day. I dropped the feed line over to the operator of the pumper, just the end of it. That is my job. Someone else is going to turn it on. They needed a ladder out back. We didn't know what was happening. I got on the back of the ladder and Frank Zwicker was on the front carrying it in. I can remember all these people standing there. They were all looking around and hollering and I am pushing the ladder forward. Poor Frank, I pretty near had him [lifted] off the ground. If you knew Frank, he might have been 110 pounds and not much more. I can remember as we came around the corner of the house, towards the back, with the ladder and we swung it around, there was a fellow standing there and I could see he was going to be in the way. I think I armed him and caught him in the chest. I didn't even realize it at the time. I was so focused on not hitting him with the ladder and getting it back there as fast as possible," Young explains.

They had to move fast.

Someone in the large crowd of onlookers told the firefighters a key fact which meant it was absolutely imperative that Young and Zwicker could not, and did not, waste even one second moving the ladder around the back of the burning home and into place. At the same time as Young and Zwicker were racing to put up the ladder, Charlene McKenzie Meade, one of several people who had fled the home, told their colleagues something that firefighters never want to hear – someone was trapped inside the burning house.

Frank Zwicker spent thirty-two years on the Halifax Fire Department before retiring in 1984. Despite the fact he is in his late eighties when I interview him in 2012, Zwicker still vividly recalls October 2, 1978.

While Young says he was uncertain of exactly what was at play initially because he was busy doing his job and running the feed line, it quickly became apparent to everyone, including the first responders, exactly what was at stake.

"There was no visible flame when we arrived, just black smoke and a lot of it, mainly inside, except it was pouring out of the windows," says Zwicker. He was, in fact, the first firefighter to run around the rear of the two-storey building with Landry. This is, by all accounts, while the lines were being hooked up by Young but before the ladder is later unhooked and carried around back by Young and Zwicker.

It is there, at the building's rear, that the firefighters are informed by someone that a baby is trapped upstairs in its crib. Zwicker says that significant piece of information is crucial to the eventual rescue because whoever went inside would have had a much better idea of exactly where to look for the child. Searching for a crib in the black smoke is somewhat easier than searching for a tiny baby who could have been anywhere inside a room or the home.

Nearby is firefighter Les Power, who plays a critical role in the rescue. His actions are essential once Young and Zwicker place the ladder in position under a window. Power, who is in his late eighties when I interview him, worked thirty years as a firefighter before retiring in August 1982.

Power says he arrived at 3390 Federal Avenue on the first emergency vehicle, #7 Engine. In the panic and chaos, Landry decides not to wait for the ladder to gain entry. Zwicker is one of many people who watched Landry start an ascent in a desperate and selfless attempt to save the child.

Zwicker confirms Landry immediately went for a trellis that formed one side of a small porch at the back door. Landry starts climbing up. When Zwicker sees Landry, he runs for the ladder. While Zwicker and Young retrieve the ladder from the rescue vehicle parked out front on Federal Avenue, behind 3390, the rescue has started to unfold.

Eyewitnesses say Landry climbed the trellis up onto a small overhanging porch roof that jutted from the rear entrance. Joe Young describes the rescue attempt from there. "Basil jumped off from the roof to his left and clung onto the gutter that ran across the roof line. He was hanging from it with both hands. He went

The back of 3390 Federal Avenue, showing the trellis (next to the drainpipe) and gutter Landry used to carry out the rescue. (Photo, taken July 13, 1979, courtesy of Halifax Police Department Identification Section.)

hand over fist, clinging onto the gutter, towards the bedroom window where the child was trapped. Basil described this all to me later because I asked him to. He said, 'Joe, I hung on with one hand and busted the window out with my helmet. Then, I went in.'"

Young explains what physical factor Landry had working for him in his favour. "It had to have been someone of Basil's slight build, with upper body strength, to be able to make that entry, coupled with a lot of will. I know if it had have been me, the gutter would have fallen off, because I would have been too heavy."

Gutter still intact, Landry somehow manages to swing and arc his body upwards and into the broken bedroom window without assistance. The mesmerized crowd of hundreds of people watch his entry. All the while, thick black smoke constantly rolls from the window. Despite the hazardous situation, Landry is not wearing any breathing apparatus when he enters the second-floor bedroom.

Bob Whorrall, who also worked out of the Bayers Road station and attended at the scene, was sixty-nine years old when I interviewed him. A twenty-seven-year veteran, he retired in 1994. He is adamant about the fact that it was very common for firefighters to not wear breathing apparatus during that era and earlier, especially those who arrived first at a scene. "We were called 'smoke-eaters.' Sometimes we would take a wet washcloth with us in our jackets and place it over our noses."

Whorrall's own report from October 2, 1978, is brief but explains his direct involvement: "Assisted No. 7 Company in extinguishment, ventilation and overhaul. Returned to quarters by order of D/C [district chief Doug Findlay]."

I asked my father about his not wearing an oxygen tank that day, why he would have taken that risk. Considering Whorrall's comments and because they all knew a baby was trapped upstairs, it makes far more sense. Dad told me, in his immediate assessment of the situation, there simply was not any time, not one second, to be wasted to stop and put on his breathing equipment. His instincts took over.

As Landry burst through the window, he hit a chest of drawers partially blocking the window, wrestled it aside and landed on the floor inside the child's bedroom amid shards of broken glass. The room was completely and utterly black – so dark he could not see his hand in front of his face. He crawled around the floor on his stomach, moving his arms out in front of him in large, arcing movements to protect himself from any unforeseen dangers and, more importantly, in order to feel the legs of the baby's crib when his hands eventually brushed up against them.

He ran his hands up the legs of the crib to its top and then gently moved them around inside the bed to locate the baby. He could not see the infant lying there but felt the tiny bundle underneath a heavy blanket, which played a role in the child's survival.

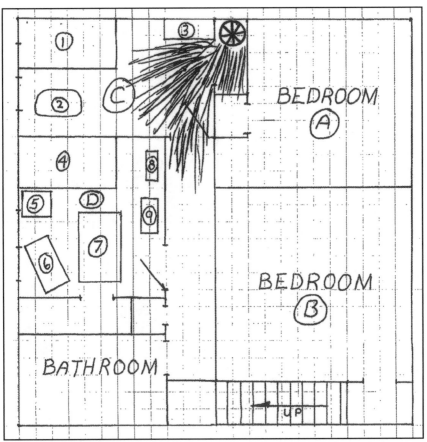

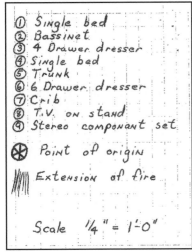

① Single bed
② Bassinet
③ 4 Drawer dresser
④ Single bed
⑤ Trunk
⑥ 6 Drawer dresser
⑦ Crib
⑧ T.V. on stand
⑨ Stereo componant set

✸ Point of origin

///// Extension of fire

Scale 1/4" = 1'-0"

Landry's diagram of the second floor layout of 3390 Federal Avenue. On the second floor, at the top of twelve stairs, is a landing and then a long, narrow hallway. The first room is the bathroom. There are four bedrooms – A, B, C and D. The baby was trapped in bedroom D. (Photo courtesy of Baz Landry's estate and Halifax Regional Fire & Emergency.)

He picked up the child, wrapped it up in the blanket to protect it from the smoke, and then, staying close to the floor, tried to find the window and much-needed oxygen for them both. At some point, he gave the semi-conscious baby mouth-to-mouth resuscitation.

As all of this is unfolding inside the tiny bedroom, several critical things are happening with the other firefighters battling the blaze on both the outside and the inside.

"While all this is going on, there is another crew in there fighting it. They were coming up the stairs. Rob Brown was there," Joe Young says.

Sixty-four-year-old Rob Brown confirms he arrived on # 7 Engine with Gordon "Champ" McIsaac and Les Power. "I probably had my back to the fire going in. I was riding on the jump seat, so you are riding backwards."

Rob Brown was the first person to make it up the twelve steps from the front of the home to the second floor. This is crucial because others who had tried so desperately in vain to make the climb, including family members, were not able to because of severe conditions. They were all forced back.

Brown was wearing breathing apparatus unlike anyone else initially at the scene. Brown explains, "The way we used to do it at the time: not everybody put breathing apparatus on. They were all in plastic cases in compartments. Most guys didn't wear them. They were relatively new [to us] at the time. If there is a situation where there is going to be a search, one person will put a breathing apparatus on – that was me on this occasion – and go into the building and try and do a search."

Not only is Brown the first person in wearing an air pack but, as a young firefighter, it is also his very first search. "I'm less than a year on the job at the time, about ten months." That day hit very close to home for the novice firefighter as he had an 18-month-old son. "I remember someone yelling that there was a child in the house. I'm pretty sure I said, 'Where?' They answered, 'Upstairs in the bedroom.'"

Brown went in, with an air pack, alone, searching. He fought through smoke and blackness and risked his life because of the baby. He says he did not initially know that Landry had already gone in after the child.

All eyewitness accounts and interviews confirm if one to two more minutes had elapsed before Landry found the baby and made it to the window for air, they both would have died. It was in that tiny room that *The Sixty Second Story* was born and Landry and the baby almost perished. Brown would have been the firefighter to recover their bodies.

He vividly remembers the scene. "By the time I got up the stairs I was crawling. You can't see anything. There is no visibility. You're not crawling because of the heat; you're crawling because you're not sure of what you are going to run into. I went into the bathroom. I think I realized it was the bathroom after I felt the toilet or the bathtub. I felt around to see if there was anything in there. There was nothing there."

Brown presses on in the blackness. "I crawled into the bedroom where the child was … I was a fair way in and I saw your dad at the window with the child giving the child mouth-to-mouth. At that time, the outside crew was bringing the ladder around and putting the ladder up."

I am floored. Until April 30, 2013, and my meeting with Brown, I had no clue whatsoever that someone else actually had witnessed what happened from *inside* the child's bedroom. I always thought, incorrectly, that Dad was the only one who made it inside.

Brown sets the scene: "I was so happy to see your father there because I know that child wouldn't have survived."

"Why?" I ask.

"They would have suffocated. I think if your dad didn't get in there when he did, he saved the child's life, there's no doubt about it, the child wouldn't have been able to breathe. Who knows how long it would have taken me to find him?"

The thought of not being able to find the boy still rolls around Brown's mind, despite these facts: thirty-five years have passed, it was a positive outcome, and everyone survived.

Brown confirms the exact timing of what happens. "It's a matter of seconds; you're not into minutes, at this point. We are talking the difference of a couple of minutes. The timeline of your dad finding him and me finding him would have been several minutes. Baz is in the building when I am getting off the pumper. It took half a minute to a minute to put on the air pack, got to go up the stairs, spend time in the bathroom, get to the bedroom, so there's a good two minutes anyway."

"When you saw a figure, did you know that figure was Baz Landry?" I ask Brown.

"Not 'til I got to him," he answers. "I'm shaking my head after I realized there was a firefighter there [by the window]. I remember saying to Baz, 'How the hell did you get in here?' I think he kind of chuckled and said something like, 'I just came in through the window.'"

Landry had managed to make it back to the window from the crib and did not get turned around or confused in the utter blackness. This is impressive because he was now suffering from an increasing lack of oxygen. Brown says, "I was standing there with him, and if he didn't have the baby I am sure I would have hugged him."

I could hug Brown. I refrain.

"He had his head out trying to get air. There was no air in the room; to hang in as long as he did and maintain his composure … I'm sure that ladder was put up quickly, but I'm also sure to him it seemed like an eternity. I told this story many times at work with younger guys, just as an example of how professional, experienced and knowledgeable some people can be."

It is also true about Brown and the entire responding crew.

The only way Landry and the baby escape the fire is because of the quick-thinking efforts of Joe Young and Frank Zwicker, who rush for the ladder when they see Landry start his ascent up the trellis. The placement of the ladder, the escape route, happens like this. Once Young and Zwicker bring it around the back of 3390 Federal Avenue, Zwicker drops the foot of the ladder down. Young hoists it up into place. It is now leaning up against the house and under the window.

"He [Zwicker] knew where it was to go, where to place it. When I pushed it up, it wasn't even up against the building and Les Power was already going up it!" recalls Young, of Power's instinct to immediately get up the ladder to help. Zwicker remembers holding the ladder in place from below as Power climbs up to the window.

Power must have initially been around another side of the building, because when he came to the back and first saw the ladder, he remembers reacting this way: "No one was at the ladder, so I thought to myself, 'Who the hell put the ladder up, and who knocked out the window?'"

Power says he rushed up the ladder thinking someone needed help. Two people did desperately need assistance. Power describes what he initially saw. "I poked my head inside the room and had to pull it right out again because it [the smoke] was so bad. I couldn't see a thing. The smoke was just rolling out of there," he says.

Power was at the top of the ladder "for about a minute" before he saw two hands reach out from the black smoke. He could not see who they were attached to, or the face. "They handed me a bundle. I didn't know what it was. I got down the ladder as fast as possible because I didn't want to waste time," says Power, confirming he is the firefighter who carried the baby down to safety.

Les Power, Frank Zwicker and Joe Young are the firefighters who gave Baz Landry and the two-month-old boy their way out. Without the ladder being in place, the outcome would have been tragic. "Everyone agrees that one to two more minutes in the black smoke and it would have been too late," concludes Zwicker.

The way the very end of the rescue unfolds at the ladder still brings driver Joe Young to tears: "When I looked up there was Baz." At this point, he becomes breathless and breaks down while explaining the amazing window scene and the descent.

I ask, "What did you see?"

"He had the baby," he says. "Baz was at the window when we went around. He was waiting for us. That's what all the people were talking about. They were looking at Baz up there. Last night we were burning wood and bush down there [he gestures to a steel barrel down by the lake in his backyard]. The flames were coming out and rolling over it and I said to my wife, 'That's what it was like the night with Basil' ... the smoke and the flames were just going out above Basil's head at the window. It was a superheated area, not a lot of flames."

At one point, as the seconds elapsed and with the ladder in place, Landry leans over the window sill while holding the baby in front of him, to protect the tiny boy from the smoke and flames. It is then he hands the baby to Power, according to Young. "I can still see Basil with that baby like it was yesterday. Les brought the baby down and then I left. As I left, someone told me that the fellow I had pushed in the chest, to safely get out of the way of the ladder, was all right. I drove him right through a fence. I was six feet one."

It is now abundantly clear there was, in fact, no time to waste getting the ladder back around and into place. Young's instinctive decision to move the bystander out of harm's way likely helped make a difference in the overall outcome. Seconds mattered.

Only when Power is climbing down the ladder does he take a second to look inside the blanket to see what was all wrapped up. Until that point, he did not know it was a baby he was carrying: "It was a bundle of joy," the firefighter recounts of his initial reaction to discovering he had just helped an infant to safety.

"Death was at its door. The blanket over its face saved it from the smoke. I was in heaven when I opened that blanket," Power says. "There was not a sound coming from the blanket. When I opened the bundle up, its eyes were wide open and it had come to. When I touched it, the baby started to cry."

And, with that, the huge crowd of onlookers erupts into wild applause and cheers. The men had just rescued two-month-old Nicholas (Nick) McKenzie.

If the ladder and crew had not been in place, Power believes Landry may have eventually dropped McKenzie out of sheer weakness, due to his heavy smoke inhalation. "Jesus, the smoke that belted out the window! I don't know how he stood it!" exclaims Power.

Of all the people I have interviewed, only Zwicker remembers my father's descent down the ladder, because he held the ladder in place while Landry climbed down to safety. Everyone else would have, of course, been focused on the baby and Power. Despite what he had just endured, Zwicker says Landry climbed down in a "regular manner." He says Landry was "dirty and exhausted."

Rob Brown, who remained inside the room with Landry until he and the baby were rescued, could not see which firefighter was at the top of the ladder because of all the smoke. "Your father went out the window ... he was calm. There was no panic. He knew it [the ladder] was coming and the guys were doing it as quick as they could ... I went back down [through the fire and out the front door]. The guys came up with a [feed] line and we started to fight the fire. It seemed to be in slow motion, but everything happened so quickly in the final analysis," concludes Brown.

Power, Zwicker and Young provided the only life-saving exit to Landry and McKenzie. "I often think I'd love to have that picture of Basil at the window with the baby," Young says.

"So would I," I quietly add.

"Well, I got it," Young says, tapping his temple with his finger, indicating that image is forever ingrained in his memory.

"I wish you could give it to me," I say. "I can tell it was very special because of the way you are reacting ... you are showing it to me the best way you can. It is a huge gift."

The best gift is that both Landry and McKenzie made it down the ladder. The Chief Duty Officer Report reveals that the police and an ambulance were also both dispatched to Federal Avenue at 18:48 (6:48 p.m.). This is eight minutes after the fire department was called. The responding firefighters took three minutes to arrive. That means this second wave of first responders raced to the scene five minutes after the fire department had already arrived and were well into the rescue.

Fire Prevention Officer Lieutenant Dave Starrett was also sent to begin a formal investigation at 19:12 (7:12 p.m.). Don Snider says Starrett became deputy chief in 1989 and passed away in 1991. Starret completed and signed his official report three days later, on October 5, 1978. The report reveals there was a total complement of thirteen firefighters who responded on the three emergency vehicles.

It also confirms the responding engines as #7 Pumper, #7 Aerial and Rescue #2.

This is Starrett's written description of the fire's point of origin and its spread: "The fire originated in the lower left hand corner of closet in bedroom C, fire extended out of this closet by burning out through the closet door. The fire then spread through the upper portion of the bedroom and burned out through the bedroom door. The fire was contained at this point." (See sketch on page 47.)

Bedroom C is the room located at the end of the upper floor hallway, on the left, just past the tiny room where the baby was. I do not have a copy of the investigator's concluding report about the fire's cause.

The fire was eventually contained by crews fighting it from the outside and by Brown and others who eventually made their way up the twelve stairs, nearer to its point of origin. There was extensive smoke and water damage throughout the home. Starrett concluded there was $10,000 to $12,000 estimated damage and loss of property, which was significant thirty-five years ago. The owner of the property was listed as The Halifax Housing Authority. Starrett described the home as "row housing – centre unit," the construction as "frame, two storey, semi-detached, row housing."

Bedroom C is also shown on the detailed, hand-drawn sketch my father created of the fire scene. Landry's drawing also confirms the fire's point of origin as coming from a closet in that bedroom, down the hall from where the rescue took place.

According to the drawing, when you enter the room where the fire originated, the closet was on the right-hand wall, farthest from the windows. Next to the closet was a four-drawer dresser. The only other items captured by Landry in his sketch are a single bed and a bassinet, which were placed nearer the windows and the far wall of the room.

Lieutenant Starrett's work also provides an accurate and complete list of all the people who were present at the time of the fire. This is how he recorded the information:

Mr. James A. McKenzie (father)
Mrs. Shirley V. McKenzie (mother)
Kim McKenzie (daughter – 19 years old)
Nicholas McKenzie (Kim's son – two months)
Nannette McKenzie (daughter – 18 years old)
Charlene Meade (daughter)
Jamie Edward McKenzie (son – 8 years old)
Shannon Meade (Charlene's daughter – four months old)
Kim Hoadley (friend of family – 16 years old)

The fire happened at the home of Nick McKenzie's grandparents, James and Shirley McKenzie.

The man who called for an ambulance for baby Nick was Captain Doug Findlay, who was acting district chief of C Platoon during 1978. Findlay served thirty-two years on the fire department and retired a year before Landry, in 1987. Findlay was eighty years old in 2012, when I interviewed him, the same age my father would have been. Findlay wrote the official Summary of Operations and Story of Fire report. I own a copy of it and so does he. The veteran was gracious enough to read it to me over the telephone when we spoke on June 4, 2012. He signed the report and forwarded it to fire chief Ron Horrocks on October 3, 1978, the day following the rescue.

An excerpt of Findlay's report reads: "While en route to #8 station by way of Northwest Arm Drive, received call via radio of a code 5-1 at 3390 Federal Avenue. While responding to this location, informed by PBX [Private Branch Exchange] that a baby was trapped in a bedroom on the second floor. On arrival, immediately transmitted a Code 5-2 and ascertained that the baby had been rescued from the second floor. I checked on the condition of the child, Michael MacKenzie [sic] age two months and found him conscious and in apparent good condition, however, radioed for an ambulance to transport him to hospital for a check-up.

"Feed lines and two 1½ pre-connect ... lines had been laid and the building laddered before my arrival. Personnel with lines on front door and up stairway were having difficulty advancing line even though they were wearing BA [breathing apparatus] ordered ventilation by removing glass from upstairs windows, backed up by an additional 1½ line in the front. When the second floor was vented, the crew working inside were able to advance line and complete extinguishment ... extensive damage to entire second floor and some water damage to first floor. Fire apparently started in a closet in a bedroom on the west side of the second floor. Called for the on duty fire prevention officer to determine cause. Lieut. Starrett arrived and started an investigation ..."

Findlay's summation to me of #7 Company's actions on October 2, 1978: "You did your best. That's all you could do ... you never knew what you could run into one minute from the next." That uncertainty can and does take a toll. Findlay's account from that time and his own conclusion about Landry's pivotal role was captured in his second report, submitted to assistant deputy fire chief Charles (Charlie) W. Robinson.

It reads, verbatim, in full:

"On October 2, 1978 at 1840 hours, a call of fire was received by Number Seven Company for 3390 Federal Avenue. While en route to this location, they were informed by PBX that a child was trapped on the second floor of this dwelling. On arrival at the scene, Lieut. B. Landry, in charge of Rescue #2, immediately determined the location of the child from parents and a neighbour who had tried in vain to rescue two month old Nicholas MacKenzie [sic] from a bedroom on the second floor. The Lieutenant ordered a ladder to gain entry by way of a window from the exterior. While waiting for the ladder, he scaled a railing and got on to the roof of a porch which was two to three feet laterally from the bedroom window where the child was located. He smashed out the window and was able to climb into the room. He grabbed the boy and brought him to the window where the ladder was then in place and passed the child out to a crew member. In my opinion, the fast thinking and quick action of Lieutenant Landry, without the aid of breathing apparatus, saved the baby's life. Because, on my arrival, conditions on the second floor were almost untenable, even with breathing apparatus."

Discussing that day with me clearly dredged up many memories for Doug Findlay because the former acting district chief called

Zwicker to chat and reminisce after I had spoken with him. Findlay also fondly remembers my father and the late captain Frank Walsh, who was off-duty sick that day, taking him deep-sea fishing years later. Findlay remembered this because he was recuperating from a hip and ankle problem. "They got me on the boat. I wasn't too mobile. They had to lug me across the jetty."

The close camaraderie was evident after the fire crew left 3390 Federal Avenue and returned to the Bayers Road station. "I can remember going back to the station and everyone felt really good about it," says Joe Young. "To me it was amazing because my role model just did something special. Just the way he [Landry] handled himself on calls. He was always cool, calm and collected. He did his job and he went beyond his job. He did not know what he would face when he went in there."

Young's conclusive comments echo Findlay's. Dealing with uncertainty is part of the package that comes with being a first responder. However, it was clear, almost from the outset, this call was for help in a desperate situation. "Someone in the driveway said, 'A baby is in there!'" Joe Young confirms, "You see, that's all Basil would have needed to hear." That someone was McKenzie's aunt, Charlene McKenzie Meade, who ran from the fire with her four-month-old daughter.

We can still read what my father had to say about that day because I have a copy of the official report he wrote as the officer in charge of Rescue #2. Landry's own words confirm it took the fire department three minutes to respond.

"While responding to stated location, #7 Engine and Rescue #2 were informed by PBX that a baby was trapped in second floor bedroom. On arrival, Number 2 Rescue crew proceeded to rear of building to carry out rescue from this location. Lieut. B. Landry was able to gain entrance to second story bedroom window to carry out rescue of Michael [sic] James McKenzie, aged two months. Child was in semi-conscious state and some mouth-to-mouth resuscitation was necessary to revive child. Child was removed to rear bedroom window and recovered by hoseman L. Power, F. Zwicker & J.

Young, who laddered building. Ambulance was ordered to transport child to V.G. Hospital ..."

Landry's report confirms the identities of the three firefighters, who are on record here and who played pivotal roles at the rear of 3390 Federal Avenue. It should be noted that my father incorrectly recorded the boy's first name, as others in the fire department and media had also done. This confusion and misreporting and/or spelling of Nicholas (Nick) McKenzie's first and last names over the years actually hindered my later search for the family. It is, however, the only inaccurate or misleading information I have come across after months of research, many interviews and the reading of at least a dozen fire department records.

Lieutenant Landry also notes that hose lines were eventually run up and into the rear centre window of the home to help aid in the extinguishment of the fire. The crew from Rescue #2 also assisted in the ventilation of the building after McKenzie was taken to hospital, as indicated by Findlay. What really strikes me about Landry's account is that it is very perfunctory in the sense that it doesn't come close to properly explaining his difficult entry and rescue without the use of breathing apparatus.

I have come to respect what my father did even more deeply because he never called attention to what happened, not in its aftermath, during the wide media coverage or later in life.

Although Landry went on to receive many accolades for his actions, he always gave credit to his fellow firefighters. I am helping to ensure #7 Company receives proper recognition with this full and detailed account. It took them three minutes to drive from the Bayers Road station to 3390 Federal Avenue. Had the emergency call or the response been delayed by one to three minutes, events at the fire scene there would most certainly have had a tragic outcome.

Baz Landry and Nick McKenzie at Federal Avenue in 1979. (Photo courtesy of Halifax Police Department Identification Section and Halifax Regional Fire & Emergency.)

Chapter 5

Awards and Accolades

Three days after the daring rescue, the accolades began.

The notification of the first one was sent to chief Ron Horrocks on October 5, 1978, by the Firefighters Commendations Board. Three senior officers signed it – deputy chief Don Swan, who was the chairman, platoon chief Cyril Ryan, and Tom Abraham, divisional officer of Fire Prevention.

It reads, in part: "Re: Actions of Lieut. Basil Landry at fire 3390 Federal Avenue on October 2, 1978. In compliance with your [Horrocks] directive of October 3, 1978, we, the Board, have examined the incident reports and are of the unanimous conclusion that commendation be of the first class: Full Commendation involving certificate with the signature of the Mayor and Fire Chief and presented at a Council meeting at City Hall. Particularly considering his grasp of the urgency of the situation, single-handedly scaling a porch roof, making an almost physically impossible entry through a window quite offset from the roof and by his own report 'finding the child in a semi-conscious state and some mouth to mouth resuscitation was necessary to revive the child.' He was not using breathing apparatus, nor was he concerned with his own physical comfort or safety, and he was, indeed, in our opinion, instrumental in the fact that the child is alive today."

I interviewed the most senior board member, retired firefighter and former chief Don Swan, on August 25, 2012. Firefighting also runs in Swan's family – his great-uncle Fred MacGillivray was department chief from 1945 to 1963, and his uncle Gerald Whalen also served in the department. Swan graciously invited me to his home to discuss the 1978 rescue, firefighting in general, and my father. Swan was deputy fire chief from 1976 to 1983, when he then took over as fire chief until December 1988. "I came to the fire department in 1955. I used to love to go to work. When I joined, it was full of fellas that had fought the Second World War and had come back. They were veterans," Swan recounts.

He was adamant during our meeting that I am welcome in his home anytime because "This is all fire department people." Every single person I have interviewed or contacted for this book has welcomed me and my inquiries with open arms. The depth and detail of *The Sixty Second Story* would not have been possible without all of them.

The former chief sets the scene of what was likely happening in Bayers Road fire station before the 1978 emergency call. "Supper was between four and five in the afternoon, typically, in the 1970s. At 6:40 p.m., when the call came in, they were usually watching a supper hour [news] show called *Gazette*. The floor watch times changed at six. Someone had to put their dress uniform on and sit at the desk in front of the station. They're in charge of security and keeping their eyes open. One of Baz's crew would have been on watch," Swan explains.

"The first thing [top priority] at any scene is 'rescue life.' Rescue life is always first and that's why you'll see people, like your father, dashing in without breathing apparatus on … Now, it was teamwork, but the fact was, Baz didn't want to wait for the ladder. He looked at the situation, sized it up and said in his head, 'I can get in there.' He just went because time is of the essence. And he was a mover; he didn't stand there as the officer and say, 'You go up there and do this.' He saw the situation. They went for the ladder and he went for the window. Everybody did their job because that's how he got out."

A veteran firefighter who fought at hundreds of fires, Swan describes what the scene inside the baby's room would have been like for Landry, without the aid of an air pack. "The fire rolls across the ceiling in little bubbles of orange and red, about the size of tennis balls. They seem to be all connected to one another. The smoke is heavy. It's black and stratified. It gets heavier the higher you go. You've got to get down as close to the floor as possible because that's the only place you can get air. Even that stings and burns. He didn't have his ears covered, so his ears would have started to tingle, burn and then blister – the ears first because they are exposed. He would feel a burning in his throat. It would have gotten very tight. Self-preservation would say, 'I've got to get out of here.' But you are looking for a child, so you keep going. It would have been agony. A human being who is trained in firefighting is going to press on. You have to try and save the baby. You have to press on no matter how much it hurts, or how much it burns!"

Swan's description is hard for me to hear. Yet it helped me further understand the pain Dad would have endured. As difficult as it was to listen to, it has made me more grateful of the outcome. Swan also explains the escape tactics and strategies my father would have used, after years of firefighting experience. "He remembered the path [of what he touched and felt]. He would be thinking as he went in, what he did, where he came, and reversed it; to know to do the opposite to get back to the window. It is like a map. You reverse the order of what you did, how you went in, when you retreat."

Swan's signature on the Firefighters Commendation Board document seals his opinion of Landry's actions. The official commendation award is among the records my father gave me. It is made of parchment and stamped with a gold seal at its bottom left. Ten days after the rescue, on Thursday, October 12, 1978, my father received it at Halifax City Council during a formal presentation. Halifax alderman Nancy Wooden presented him with the document for bravery. A photograph of Wooden handing Landry the award, with Ron Horrocks looking on in the background, appeared the following day on the front page of *The Mail Star*. I have the complete edition of the now yellowing newspaper which my father also kept. The story, written without a byline, is accompanied by a large photo and appears below the fold line, in the lower right-hand corner of the paper. It discusses the rescue and explains that Landry had just received the fire department's highest award for bravery. Mayor Edmund Morris requested that a record of the presentation be placed in Landry's personal staff employment record on file.

Interestingly, another article about firefighters appears in the same issue. On the flipside of the front page, on the lower left of page two, there is a shorter story, without byline or photo, with this headline and description:

FIREFIGHTERS WILL TREAT YOUNGSTERS

Halifax firemen will conduct a day for young people on Saturday at the Wanderers Grounds starting at 1:30 p.m. Children 12 years of age and under will be treated to rides on a fire truck and free pop and chocolate bars. The firemen will be on hand to answer any questions the children might wish to ask, and to give out pamphlets on fire prevention to children and their parents. In an effort to make the younger generation more fire safety conscious, some of the equipment used in fire prevention will be on display. The day's activities are part of National Fire Prevention Week.

That tiny article reveals that Landry received the award during a key week for firefighters nationally.

Landry's recognition grew in scope. The following year, in 1979, as news and word of the rescue spread, the forty-six-year-old was named Canadian Firefighter of the Year. That meant Landry was forty-five when he rescued McKenzie. My father had certainly been in good shape for a middle-aged man when he undertook the physically challenging entry and rescue. He remained slight of build and in relatively good health until his heart and breathing problems began, in earnest, in 2005.

Former chief Ron Horrocks confirms he had been doing some work for Fireman's Fund Insurance Co. of Canada in 1978 when he became aware of its relatively new but prestigious Canadian Firefighter of the Year award. It was Horrocks who nominated Landry for the national honour. A copy of the nomination form, filled out by Horrocks, reveals it was submitted by the chief on June 4, 1979, to Mr. Emile Therein, executive director of the Canadian Association of Fire Chiefs, in Ottawa. The facts on the form reveal Landry was a twenty-two-year veteran of the Halifax Fire Department when the 1978 rescue occurred.

"It was a big deal. It was a very big deal," says Ron Horrocks. "The thing that comes to me is pride. I was so proud of him; proud of him carrying out the job." I spoke with the seventy-eight-year-old former chief on May 4, 2012. He served thirty-eight years on two fire departments in Canada, seven of them as chief of the Halifax department from 1976 to 1983, before retiring in 1994.

The former chief wrote a memorandum about Landry's national award in his briefing submitted to Halifax City Council and mayor Edmund Morris on July 11, 1979. In his write-up Horrocks says, "I am proud to inform you that Lieutenant Basil B. Landry, for his daring rescue ... has been selected by the Fireman's Fund Insurance Company of Canada as being deserving of the second annual Canadian Firefighter of the Year Bravery Award. This award will be presented to Lieutenant Landry at the annual banquet to be held on Wednesday, August 22, 1979, during the 71st annual conference of the Canadian Association of Fire Chiefs, which, this year, is being held at the Hotel Nova Scotian in Halifax."

Landry also received a congratulatory note from Mayor Morris. It is handwritten on letterhead from the Office of the Mayor of Halifax and is dated July 12, 1979. It reads, in part: "Dear Basil: ... the award, while singular to yourself, is a reflected honour upon the Halifax Fire Department and upon the City itself, and I join in sending you our warmest congratulations and appreciation, in the best tradition of an outstanding Departmental Force."

When the award was formally presented to Landry, I was fourteen and in junior high. My parents attended the event, where my father was presented with an engraved plaque, which still proudly hangs in my mother's home. He also received a cheque for five thousand dollars. It was a significant amount of money for our family, given the salaries of firefighters in the 1970s.

My father used part of the award money to take a vacation in the late summer of 1979. I accompanied him on a trip between Nova Scotia and Maine. We drove to Yarmouth and boarded the MS *Caribe,* which served from 1976 to 1981 as a summer car ferry between Portland, Maine, and Yarmouth, Nova Scotia. During its era, *Caribe* also cruised near Estonia but spent most of its time plying the waters of the Caribbean, before ceasing operations in September of 2011.

My mother did not accompany us as she has a fear of open-ocean sailing. That 1979 trip remains one of my favourites. It was my first time outside of Nova Scotia and Canada. I fondly remember being on the ship, and after landing in Maine, eating at a Denny's restaurant. As a teenager I thought it was cool to eat at the popular American food chain. Afterwards, we went shopping and walking along the famous boardwalk at Old Orchard Beach. It was a wonderful time to share with my father. I still smile whenever I am vacationing in the United States and drive past a Denny's.

My father and I had one final opportunity to sail via cruise ship, in 2003, on the occasion of his seventieth birthday, three years before his death. We flew from Halifax to Florida and departed from Port Canaveral to the Bahamas on a five-day cruise. Once Royal Caribbean staff found out he was celebrating a milestone birthday, he was treated like royalty. We were even chosen to dine with the captain during a formal dinner and later toured the bridge, which was a privilege, considering the many new and stiff safety regulations and laws surrounding travel after the horrific events of 9/11.

A 1979 media release issued by Fireman's Fund Insurance reveals some of my father's other lifelong passions, besides his family, the ocean and ships: "Lieutenant Basil B. Landry and his wife Theresa celebrated their silver wedding anniversary [25th] earlier this year and they live with their 14-year-old daughter Janice ... He spends much of his spare time in the summer perfecting his golf game. He has a handicap of 10 and is a member of the Brightwood Golf and Country Club at Dartmouth, N.S. He is also a keen deep sea fisherman."

A second press release confirms Fireman's Fund had only recently announced the creation of the prestigious national award on April 5, 1978, just six months before the McKenzie rescue happened. It underlines several important points about first responders: "As a company in the business of insuring dwellings, commercial and industrial plants and buildings, Fireman's Fund Insurance Company of Canada is particularly aware of what Canada's firefighters are doing daily to save and protect lives and property. The purpose of the Award is to provide recognition and a tangible reward for Canada's fire departments and making Canadians aware of the heroism that is part of the daily lives of Canada's firefighters. It is hoped that Canadians will learn to appreciate that every time they hear a fire engine's siren, a group of firefighters may be on their way to risk their lives to save someone else's. The Canadian firefighter of the inaugural year, 1978 ... was a 14-year-veteran of the Toronto Fire Department, 36-year-old William Davies, for his actions during a rescue in a house fire on December 30, 1977."

A senior representative from Fireman's Fund, Walter Atkinson, further discussed the debt owed to all firefighters during his speech in Halifax, at the hotel where the award was presented to Landry. Atkinson said, in part, to the assembled crowd, "No matter who we are or where we live, every Canadian owes a debt to Canada's firefighters who are ready, every day of the week, to risk life and limb to protect our property and, if necessary, save our lives. Even though the vast majority of the public will never find themselves in a situation where survival could depend on the professional skill and devotion of a firefighter, nevertheless, the debt is there."

A media story about the national honour appeared in *The Mail Star*, written by staff reporter Elissa Barnard. It was accompanied by a large photo, depicting Landry on the right holding the award and on the left Kim Gibson; between them is her 13-month-old son Nick.

While they were not reunited later in life, Gibson was invited to the event where my father received the Canadian Firefighter of the Year award. Gibson remembers being terrified about speaking to the media. She says she felt unable to properly describe what the rescue meant to her and her family. She also says she did not know how to appropriately thank Landry, given the magnitude of what he had done. Gibson's consideration and willingness to revisit with me one of the most terrifying days of her life is certainly thanks enough for my family and me.

Barnard interviewed Landry for her 1979 story. It reads in part, "... Lieutenant Landry said he only had time to think of the mechanics while rescuing Nicholas and noted that, 'It's not always the individual which counts, it's your total responding company. I am extremely honoured to receive the award not only on my own behalf but on behalf of all firemen.'" When he had a public platform, the only time in his life, Landry gave his peers the credit.

Landry gave another interview, this one with Lorne Campbell, the editor of *The Canadian Firefighter Magazine.* Campbell had been in Halifax to attend the fire chiefs' convention and to meet and interview Landry. In Campbell's story in the September/October 1979 issue, he wrote a lengthy Editor's Note which appears after it, recounting his personal meeting and interview with my father. It adds details about what it was like inside McKenzie's bedroom.

"At this point he [Landry] was in for a shock, just inside and blocking his entry was a dresser. With his free hand, he shoved the dresser out of the way and with his strength going fast, he pulled himself through the little window. He lay on the floor of the baby's room and said that he could breathe because of air space about a foot off the floor ... After listening to Basil Landry relate his story to me, I can't help but say how proud that we all should be of one of our Bravest of the Brave."

Shortly after, Landry was also recognized south of the border by an American publication, *Firehouse Magazine,* based in New York. He received a one-hundred-dollar cheque and a special certificate for Heroism and Community Service. The award came as a surprise to Landry, as he told *The Mail Star* reporter Peggy Miller in 1979, "After winning the Canadian award, your name usually goes into newspapers and firefighting publications. This is obviously how the American magazine heard about me."

Miller wrote about the fact that Landry's Canadian award was only the second of its kind that *Firehouse* had ever given out. "This year the magazine is giving away $10,000 to firefighters across America and Canada who have distinguished themselves in some special way through acts of courage ...," Miller said. "Editors of the magazine feel the special awards program should help to convey real appreciation to firefighters whose courage, too often, goes unrecognized."

The panel of judges cited on the New York award was made up of Gordon Vickery, administrator of the United States Fire Administration; James Shern, 1978-1979 president of the International Association of Fire Chiefs; and Frank Palumbo, secretary of the International Association of Firefighters.

It was clear with the American award that Landry's act and story was now spreading and reaching the upper echelons of fire services across North America. As a result, Landry also received many congratulatory letters and cards. He kept them all. They came from across the Maritimes, Canada and the United States. One of these, dated November 4, 1979, came from J. David Beck, the chief of the Lunenburg, Nova Scotia, fire department, who was also president of the Maritime Fire Chiefs' Association.

Beck wrote to Landry: "While we read and hear about these courageous actions happening throughout this country of ours, it is not very often that we hear of it happening right on our door step, however, we all know in our own minds that, if the occasion should arise, we have the men with the courage and capabilities to take the necessary action to save lives and prevent loss of property, right here in the Maritime Provinces. You have, without a doubt, proved that."

The strategic role of our first responders is discussed in publications like *An Historical Celebration, 225 Years of Firefighting in Halifax, 1768-1993*. Its readers get a better sense of the dangerous nature of the profession through detailed descriptions of a number of serious fires fought by the Halifax Fire Department. During 1978-1979, a number of devastating Halifax fires claimed lives and caused injury and loss of property. Five of the fire calls read:

"Fire claims 5 lives during 1978, and 4 in 1976, with none in 1977. Seven of the nine deaths are caused by careless smoking. ... Fire completely destroys Dalhousie Rink on May 14th. The damage from this fire is in excess of $1 million and several firefighters are injured, one of them seriously. On May 27th fire destroys the C.M.H.C. office building on Gottingen Street. ... LaScala Restaurant, Dresden Row is gutted by fire on July 23rd. ... Fire breaks out in a high-rise apartment at 210 Willett St. on January 28th, 1979. An aerial truck and snorkel are used to rescue residents from upper floors. A Halifax landmark, Brunswick Street United Church is completely destroyed in a $1 million fire on June 30, 1979."

The five lives lost in 1978 easily could have been seven as a result of the extremely close call at 3390 Federal Avenue.

The fire at the Brunswick Street United Church is again referred to in the Halifax Fire Department's Annual Report of 1979. My father is pictured on its front cover, wearing his dress uniform and standing in front of Rescue #2, on which he drove, with Joe Young behind the wheel, to rescue McKenzie. Unusually, Rescue#2 is white and not red in colour as are most fire engines. Because of its unusual appearance, several firefighters told me it was nicknamed the Ice Cream Wagon or the Ice Cream Truck.

At the conclusion of the annual report is a selection of Letters of Commendation highlighting the work of various Halifax firefighters and companies, including those who responded to the church fire. A letter was sent by a group of residents to chief Ron Horrocks on July 13, 1979. It partially reads:

"On the evening of Wednesday, June 20, 1979, Brunswick Street suffered a major calamity when the Brunswick Street United Church was destroyed by fire. As the owners and residents of 2125 Brunswick Street, the house immediately adjacent to the church, we experienced a special involvement in this fire. Distressed and distracted though we were by the evening's events, we could not fail to notice and appreciate the courage and care that was exhibited by the firefighters who were battling the blaze. They performed a truly excellent job of containing the fire. Although we were evacuated from our home, as were several of our neighbours, damage outside

Baz Landry in dress uniform beside Rescue #2, affectionately referred to as the Ice Cream Wagon or the Ice Cream Truck because of its white colour. (Photo courtesy of Halifax Police Identification Section and Halifax Fire & Emergency.)

the boundaries of the church was minimal. The firefighters are to be deeply and most gratefully thanked for this. In closing, let me again say that we found the city's firefighters to be courteous, cheerful, hardworking, and courageous ..."

The letter is signed by eight people.

Further praise for Halifax firefighters appears in another commendation letter, handwritten and dated February 20, 1979. It was sent to the Lady Hammond Road station, where the monument to fallen line-of-duty firefighters now exists. It says, "On Sunday, Feb. 18, 1979, your firemen answered a call to our home. We would like to comment on their excellent performance. They were very efficient and professional. What impressed us most was the apparent ease and confidence with which they worked. Also they seemed to make a real effort to make as little mess as possible. Our fears of 'mass destruction' never materialized as they covered the carpet and even finished off with broom and shovel. Their respect for personal property is a great asset to our community." The letter was signed by Diane and Greg Pottie.

Concern for property is crucial but helping people is paramount, as retired chief Don Swan has explained. That is clearly reflected in this next letter, which was sent following a life-and-death situation that occurred on the waterfront. The letter recounts the events of the day and was sent by A. Taylor, then director of police and security for the Port of Halifax:

"It has come to my attention that members of your Department were called to Pier 29 during the early morning hours of Saturday, September 22, 1979. An intoxicated ... seaman had fallen off the wharf into the Harbour waters below. Our members, not being in possession of a ladder, were unable to rescue this individual who was clinging to the rubber fenders below. Upon the arrival of your Department, a ladder was placed in position. Lieut. Tim Bookholt of your Department went down the ladder and placed a safety line on this person. With the help of others at the scene, police, C.N.R. employees etc., this individual was pulled to safety and then taken to the V.G. Hospital, where he is presently recuperating. I would like to take this opportunity of thanking those members of your

Department who so capably assisted, in particular, Lieut. Bookholt … Undoubtedly, their prompt action saved this man's life."

Another example, which happened at National Sea Products Ltd., was described in a thank-you note sent by division manager George Phalen in the spring of 1979. Phalen gave due credit to platoon chief Cyril Ryan and his crew. Ryan was one of the three men who signed the commendation board letter for my father.

Phalen's correspondence was also selected to become part of the Letters of Commendation showcased in the annual report created by Ron Horrocks: "On May 17, 1979, I had a gaseous condition of unknown origin in the warehouse on Terminal Road. I requested assistance from your department to investigate the nature and cause of the gas and received an immediate response from Platoon Chief Cyril Ryan, his men and equipment. They determined a chemical, Oakite, used for cleaning metals and descaling boilers, was creating the gas. It is comforting to know that there are highly trained men available with the ability to deal effectively with emergent situations … I would deem it a personal favour if you would kindly convey my appreciation and thanks to Platoon Chief Cyril Ryan and his men for their valued assistance."

These are just a few of the numerous accolades given to many Halifax firefighters during 1979. However, a record-setting milestone, stemming from the 1978 McKenzie rescue, had yet to unfold.

Chapter 6

Medal of Bravery

The year 1980 began with tragedy for the Halifax Fire Department. A fire broke out on January 20 at 2117-19 Gottingen Street, involving several stores and apartments. During that emergency, firefighter Allen MacFarlane, age fifty-one, collapsed at the scene after being inside the building and suffering severe smoke inhalation. MacFarlane was taken to hospital but died afterwards. He was the platoon chief's aide at West Street fire station.

MacFarlane's name is officially recorded within the fire department as a line-of-duty death. His name is also engraved on the fallen firefighters' monument. As of this book's writing, during 2012-2013, MacFarlane is the last recorded firefighter death in the history of Halifax Regional Fire & Emergency.

While deep sorrow permeated the fire department at the start of 1980, an announcement in the summer helped, at least in a small way, to ease the pain. Lieutenant Baz Landry was officially notified he would receive the Medal of Bravery (M.B.) from the Canadian government, at Ottawa's historic Rideau Hall.

My father received a formal notification letter, dated July 21, 1980. It was typed on Rideau Hall letterhead and sent from Roger de C. Nantel, the director of the Chancellery of Canadian Orders and Decorations: "I am writing to let you know that His Excellency, the Governor General, today approved the award to you of the Medal of Bravery in recognition of your action at Halifax, Nova Scotia, on the 2nd of October 1978. The Medal of Bravery is part of the Canadian system of honours and was instituted in 1972 by Her Majesty the Queen on the advice of the Canadian Government. An Investiture will be arranged for recipients of bravery awards and when the details are completed, you will be invited to come to Government House to receive your decoration. Your award will be announced in the press and published officially in the next *Canada Gazette*. You are now entitled to use the initials 'M.B.' after your name, when appropriate. The Governor General has asked me to extend to you his warm congratulations, to which I would like to add my own."

My father never used the letters M.B. after his name, nor did he ever really speak about this honour. In fact, he said several times to me that he never thought what happened in 1978 was particularly noteworthy – it was part of the job. While he was truly honoured by the medal, he always downplayed it because of his nature.

However, the medal being awarded to Landry is significant and historic; it represents a first for Halifax Regional Fire & Emergency. As of mid 2013, when I completed writing, Landry remains the sole recipient of the Medal of Bravery within the Halifax fire service. Keep in mind, Halifax has the oldest fire service on record in Canada.

I have triple-checked the accuracy of this achievement with three reliable and respected sources: from firefighting historian Don Snider, through the current fire chief's office, and finally through several communications with Afton Doubleday, constituency assistant in the office of Geoff Regan, Member of Parliament for Halifax West. Doubleday has been an integral part of fact-checking the Ottawa end of this story.

I carry a worn snapshot of my father at Rideau Hall in my wallet. It is dog-eared and fading from many viewings. It shows both my parents, in their forties, at Government House. They both look radiantly happy.

That day was Friday, September 26, 1980.

Fourteen Canadians received the Medal of Bravery that day. Here is the verbatim account from the official investiture program which was presented to my father and all ceremony attendees and medal recipients:

Lieutenant Basil Bernard Landry

Acting well beyond the call of duty, Basil Landry, a Halifax fireman, saved two-month-old Nicholas MacKenzie [sic] from a fire in his home on 2 October 1978. From outside the house, the baby's upstairs bedroom was pointed out to Lieutenant Landry, who climbed a trellis leading to a veranda situated below and to the side of the bedroom window. Smashing the window, he pulled himself into the room. Being without his air-pack, he stayed close to the floor and made his way to the crib. After giving the baby mouth-to mouth resuscitation, he carried him to safety.

Along with the program, which briefly described the brave acts of all recipients, the group was also presented with a second booklet, dated January 1981, detailing the complete list of every recipient of all three of Canada's decorations for bravery – the Cross of Valour, the Star of Courage and the Medal of Bravery. As of that date, more than 300 Canadians had received the Medal of Bravery since its inception into the Canadian Honours System in 1972.

Four days before the ceremony, my father received a personal note, on official letterhead, from Canada's Leader of the Opposition, Joe Clark: "The Medal of Bravery is a well-deserved tribute to your valour and dedication and I know the people of Canada look upon you with respect and admiration."

Progressive Conservative Member of Parliament, the late Howard Crosby, who represented Halifax West, our family's constituency, was present for the ceremony. Crosby, who served his riding for fifteen years, died at age seventy in 2003. However, we can trace his thoughts in an excerpt from his *Constituency Report* circulated at the time. It reads, in part: "Halifax West has a real life hero! I was present when the Governor General of Canada presented the award at Government House in Ottawa. Let's remember all those who risk their lives to save and protect others …"

Governor General Ed Schreyer presented the fourteen recipients with their medals. The Governor General of Canada's official website explains Schreyer's transition into the role: "One of Edward Schreyer's first encounters with Rideau Hall came in 1975 when he was awarded the Governor General Vanier Award as an Outstanding Young Canadian of the Year. Three years later, Mr. Schreyer was appointed Governor General and he and his family moved from Manitoba into Rideau Hall. At 43, Edward Schreyer was the youngest Governor General since Lord Lorne in 1878 (33 years old) and Lord Lansdowne in 1883 (38 years old)."

The purpose of the Medal of Bravery is explained in detail on the Governor General's website: "The Medal recognizes acts of bravery in hazardous circumstances. It is a circular, silver medal: on the obverse of which is a maple leaf surrounded by a wreath of laurel, and on the reverse of which appear the Royal Cipher and Crown and the inscription BRAVERY BRAVOURE. Decorations for Bravery recognize people who risked their lives to try to save or protect another. From entering burning buildings to calming gunmen to plunging into icy waters, the recipients put their lives on the line to help another person … The degree of risk faced and persistence despite the risk are important in the evaluation. In addition, perception of risk is a factor – people who try to help, even though they know they might be severely injured or killed, display bravery of a very high order. The medal is sanctioned by the Canadian Government to only be worn, by male recipients, on the left breast, suspended on a ribbon of red with three blue stripes. There is also an official medal-wearing position for female recipients; below the left shoulder, suspended on the same ribbon, made into a bow."

My father's medal rests in a lovely blue, leather-bound box, lined in dark blue velvet, with the initials M.B. engraved in gold embossing on the front of the box. The medal hangs from the very ribbon described above. I never saw my father wear it once. It was always kept this way, hidden away, but obviously cherished. It is a treasured possession.

Governor General Ed Schreyer pins the Medal of Bravery on Baz Landry, September 26, 1980.

I am honoured my father entrusted it to me, given what he had to do to achieve it and because so few Canadians are presented with one. This is the most prestigious award given to any Halifax firefighter, as of this writing.

My father discussed his medal notification with *The Mail Star* reporter Marilyn MacDonald during July 1980. The following is the excerpt that convinced me to write *The Sixty Second Story*. The sentiment expressed by Landry is an accurate representation of what my father thought of his firefighting comrades and what happened on October 2, 1978: "Lieutenant Landry, who has been a fireman in Halifax for 24 years, said he views the honours as tributes to all firemen because saving a life, 'is not an individual effort … you go to a fire as a team.'"

"You go to a fire as a team" – those eight words have reso- nated with me for years. They have been the catalyst for two years of research, interviews and writing. As much as I am proud of my father's medal, I am equally, if not more, proud of the fact he thought this way, had respect for his fellow firefighters and always approached firefighting with a team attitude.

I was attending junior high at the time of the ceremony and did not accompany my parents to Ottawa. I am not sure if it was because of my age and the fact I was attending school, or because the cost was prohibitive. My mother says she believes it was the latter.

While they were away, I stayed overnight at a friend's house and we watched a news story about the medal ceremony on tele- vision. I vividly remember being in their living room, with tears streaming down my face, as I saw my father standing at attention, wearing his dress uniform, in the ornate room.

Additionally, yet significantly, Laura is also in junior high and the same age I was, thirteen, when her grandfather rescued McKenzie. Now that I see where Laura is in her development, from a mother's perspective, it has become more alarming to me that I almost lost my father when I was that age. My life path most cer- tainly would have been far different if those extra sixty seconds had elapsed.

According to the Governor General's website, anyone can nom- inate a deserving individual for a Medal of Bravery: "Nominations must be made within two years of the incident, or within two years after a court or a coroner has concluded its review on the circum- stances surrounding the incident or act of bravery. The Police Ser- vices investigate eligible cases to ensure that information is accurate. … Nominations are received by the Chancellery for review by an independent advisory committee … The Advisory Committee is made up of representatives of the Clerk of the Privy Council, the Office of the Secretary to the Governor General, the Commission- er of the Royal Canadian Mounted Police, and the Deputy Minis- ters of Canadian Heritage, the Department of National Defence and Transport Canada, as well as up to four others appointed by the Governor General."

You can access the complete Medal of Bravery list for Nova Scotia or the rest of Canada by visiting the official website of the Governor General at www.gg.ca. Once on the site, look for the "Honours" drop down bar at the top of the homepage. After clicking on it, you will see below it a box including "Find a Recipient". That is the link you need to access all of the Medal of Bravery recipients for Canada.

Afton Doubleday, in M.P. Geoff Regan's office, sent me the complete list of Nova Scotians who received the Medal of Bravery. The document reveals that as of 2013, 316 Nova Scotians have received this high honour. My father's name is on the thirteenth page e-mailed to me by Doubleday, the 157th name listed, about halfway through the impressive list.

I am sharing a small sampling of those heroic acts to underline the valour of our citizens and the importance of the Canadian Honours System. Some of the people involved are first responders and some are not. The list I received is alphabetical, not by date of presentation. To show gratitude for all mentioned, and to try to bookend the list, I am recording the official details of the first and last person named, both of whom are male recipients. It should be noted, however, that eleven of the recipients are women. Therefore, to balance the reporting, I am also including the synopsis of the first and last females who appear alphabetically.

The citation for Fraser Agnew highlights the heroism of many people during a tragic event that occurred on May 9, 1992, in Nova Scotia when an underground coal mine exploded. It plunged families and a province into mourning and garnered world-wide media attention:

Fraser Agnew, M.B., New Glasgow, Nova Scotia
Medal of Bravery
Date of Instrument: September 8, 1994 (notification/decision date)
Date of Presentation: November 28, 1994
After a massive explosion inside the Westray Coal Mine at Plymouth, Nova Scotia, on May 9, 1992, two hundred and one men attempted to

rescue twenty-six miners trapped underground. For the next five days, rescue teams searched around the clock for survivors. Despite the perilous mixture of highly volatile gases, they repeatedly went underground, beneath a largely unsupported and unstable roof structure. They crawled around twisted pieces of steel that had once been supporting arches, passed machinery that was no longer useful or recognizable, climbed over rockfalls and debris, and waded through brackish water. The teams moved cautiously and silently, fearing that the slightest noise or vibration would cause further rockfalls or explosions. Rescue efforts were finally abandoned after more than a kilometre of the mine had been searched, with no signs of survivors. The rescuers recovered the bodies of fifteen of the twenty-six victims.

It is important to note that it was a very large team of people who selflessly searched for friends, miners and loved ones despite the extremely hazardous conditions. In fact, a House of Commons transcript, coupled with communication with Ottawa, through Afton Doubleday, confirms a total of 201 people, including Agnew, received the Medal of Bravery for their efforts during this rescue effort. Most, but not all, are Nova Scotians. Two medal ceremonies were performed, and some of the honours were awarded at individual presentations. Sixty-four percent of Canadians who had received the Medal of Bravery as of early 2013 were involved in the rescue effort in the aftermath of the Westray Explosion.

A portion of the medals was distributed and awarded in the largest ever civilian investiture ceremony, by the late Governor General Ramon (Ray) Hnatyshyn in November 1994 at Sharon St. John United Church in Stellarton, not far from where the explosion occurred.

In an article written in 1995, a year after the landmark ceremony, Cathy Hallessey, communications director with the Communications, Energy and Paperworkers Union of Canada, describes the emotional story of another of the 201 rescue workers: Byrne MacIntyre, M.B.: "Byrne MacIntyre will never forget the events of May 9, 1992. The President of the CEPU Local 823 and member of the Windsor Salt Mines Draeger Team, from Pugwash, Nova Scotia, went underground … in an attempt to rescue 26 miners … The heroic efforts of Brother MacIntyre and seven other men from the

Canadian Salt Company have since been rewarded by Canada's Governor General Ramon Hnatyshyn in an emotional ceremony that brought together many of the brave and bereaved who share the tragic memories of the Westray Mine. 'All of them, without hesitation, entered the mine prepared to risk their lives for their fellow miners,' said Hnatyshyn, in a church packed with miners, their families and the families of the 26 dead. 'In a five-day ordeal, they performed incredible feats of strength and endurance in the face of perilous conditions,' he added. Each of the 177 men to receive medals were called to the front of the church one at a time. Some faces beamed with pride, others were close to tears. This honour makes them part of the largest group in the history of the Canadian Honours System ... to be awarded bravery decorations for a single incident. 'I'll never forget it,' said Byrne MacIntyre, in a moment of pride tinged with pain."

Another group, albeit this one much smaller, is also the focus of the events surrounding the last male name listed alphabetically in the Nova Scotia portion of Medal of Bravery recipients.

Wayne Zacharuk, M.B., of Middleton, Nova Scotia
Medal of Bravery
Date of Instrument: April 3, 1997 (notification/decision date)
Date of Presentation: December 5, 1997
James Richard Stuart Brennan, M.B.
Sergeant Michael James Langdon, M.B., C.D.
Sergeant Wayne Walter Zacharuk, M.B., C.D.

On May 26, 1996, James Brennan and off-duty Sgts. Langdon and Zacharuk rescued two canoeists from Ramsey Lake, Annapolis County, Nova Scotia. While fishing from shore, the men witnessed a canoe tip over and spill two fishermen into the freezing waters, 50 metres out. First to jump into the lake, Mr. Brennan and Sgt. Langdon swam to the man who had started to submerge. The panicked non-swimmer dragged Sgt. Langdon under repeatedly before Mr. Brennan was able to get a firm grip on him. Exhausted, and in trouble himself, Sgt. Langdon then returned to shallower waters. In his continued struggle, the man pulled Mr. Brennan under several times before losing consciousness. Mr. Brennan towed the limp body to shore where the man started to breathe again. In the meantime, Sgt. Zacharuk had gone to the rescue of the other man who, by then, had disappeared under the surface. He found the man nearly six metres deep in a sitting position, grabbed him by the collar and pulled him to the surface. Sgt. Zacharuk dragged the victim toward shore until Sgt. Langdon arrived to help. Together, they returned to shore where the man regained consciousness."

Appearing seventh on the list is the first female name.

Mona Jean Balcom, M.B., of Annapolis County, Nova Scotia
Medal of Bravery
Date of Instrument: February 20, 1984 (notification/decision date)
Date of Presentation: July 9, 1984

Very early on the morning of 4 June 1983, Mona Jean Balcom rescued a ten-year-old girl from almost certain death. While driving to work, Mrs. Balcom saw smoke rising from the roof of a house. She stopped, ran into the house and found a mother with her daughter; as she urged them outside, the frantic mother screamed that her other four children were trapped upstairs. Thick smoke from the stairwell forced Mrs. Balcom to seek an alternative means of rescue. Outside, she seized a ladder and placed it against the east wall of the house, then climbed, smashed a window and, through the billowing smoke, shouted to the children. When onlookers said that a young girl had been seen through another window, she carried the ladder to the front of the house and again went up and smashed the window. While flames were engulfing the room, Mrs. Balcom stretched her head and shoulders through the broken window and pulled out the girl. Sadly, the other three children perished.

The final woman appearing on the list, as of early 2013, died during her rescue efforts.

Anastasia Jileen Sylliboy (posthumous), Eskasoni, Nova Scotia
Medal of Bravery
Date of Instrument: February 28, 2000 (notification/decision date)
Date of Presentation: December 8, 2000

On 15 July 1999, Tonia Sylliboy, her father Maxim and his cousin's daughter Anastasia combined their efforts to save two boys, seven and eleven, who had been caught in the current of a channel at Castle Bay, on Bras d'Or Lake, Nova Scotia. When he heard the boys' screams, Max Sylliboy battled the turbulent waters until he reached the young victims, some 40 metres out. Seeing him struggle to keep the children afloat, nineteen-year-old

Anastasia went to help and the panicked boys repeatedly pulled both rescuers under. A witness who was five months pregnant at the time, Tonia, grabbed a float and made her way to the scene. Tonia saw both her father and Anastasia disappear under the surface but managed to reach one boy, then the other, and all three drifted to safety, holding onto the flotation device.

Anastasia Jileen Sylliboy, M.B., and Maxim Bernard Sylliboy, M.B., both received the Medal of Bravery posthumously. Tonia Elizabeth Sylliboy, M.B., the surviving rescuer, also received one. This is a heartbreaking account and an example of the outstanding bravery of a group of incredible Nova Scotians who risked everything for others; two paid the ultimate price for their valour. Tonia, who was pregnant, pressed on towards the children after seeing her loved ones disappear right in front of her eyes.

Another group of rescuers forged ahead, like Tonia, to make a daring rescue at sea that led to four of them receiving the Medal of Bravery in 2013; they are the most recent Nova Scotia recipients.

I noticed an online story about the four during the last week of my research before I handed the manuscript for this book over to Lesley Choyce at Pottersfield Press. The timing of this latest group receiving their medals is perfect; it underlines the continued importance Canada places on the award for bravery, and, by coming near the end of my lengthy research, it has been encouraging. This latest group is from the Canadian Coast Guard. The following is an excerpt from John DeMings' story published on February 8, 2013, in *The Digby Courier*:

"Ian McBride of Digby is among members of the Coast Guard station in Westport being awarded the Medal of Bravery today in Ottawa for the … rescue of three men whose fishing boat sunk during a storm in the Bay of Fundy. The decoration for bravery is being presented this afternoon in Rideau Hall by Gov.-Gen. David Johnston."

A media release issued by the Governor General's office listed three other Coast Guard crew members who also received the Medal of Bravery stemming from the same rescue. This is the verbatim citation from the Canadian government:

Dale Burton Bollivar, M.B., Middlewood, Nova Scotia
Ian Frederick McBride, M.B., Digby, Nova Scotia
Paul Alexander Oliver, M.B., Annapolis Royal, Nova Scotia
Wayne David Pink, M.B., Albert Bridge, Nova Scotia

On December 16, 2009, Canadian Coast Guard officers David Pink, Paul Oliver, Dale Bollivar and Ian McBride rescued three men whose fishing boat had sunk during a storm in the Bay of Fundy. Tasked with the mission, Commanding Officer Pink and his crew made their way with great difficulty through troubled seas to reach the victims, who were spotted in an inflatable lifeboat. Once the victims were safely on board, the rescuers continued to search for a fourth man over several more hours, in the crashing waves. Sadly, he was never located.

According to the report in the Digby newspaper, as of 2013, Pink was stationed in Louisbourg, while McBride, Bollivar, and Oliver remain members of the Westport crew. The four Coast Guard workers were among forty-six Canadians to receive the Medal of Bravery on February 8, 2013.

Four Stars of Courage were also awarded. The media release aptly explains the rank or order of the three decorations for bravery, within the Canadian Honours System: "The three levels of the Decorations for Bravery reflect the degree to which the recipients put themselves at risk: The Cross of Valour (C.V.) recognizes acts of the most conspicuous courage in circumstances of extreme peril. The Star of Courage (S.C.) recognizes acts of conspicuous courage in circumstances of great peril. The Medal of Bravery (M.B.) recognizes acts of bravery in hazardous circumstances."

Basil Landry, M.B., is certainly in good company.

Chapter 7

The Toll it Takes: PTSD and CIS

When it is dark enough, you can see the stars.
– Ralph Waldo Emerson

First responders face countless dramatic, tragic and/or life-altering situations over the course of their careers. The next two chapters focus on the long-term effects and impact, both physical and emotional, of working in a job where you constantly put the needs of other people before your own.

In the decades that my father worked as a firefighter, beginning in the 1950s, not much was known about these issues. Terms like post-traumatic stress disorder (PTSD) and critical incident stress (CIS) had not yet been coined, let alone discussed or understood.

The current Halifax Regional Fire & Emergency (HRFE) now has a dedicated staff member who helps employees and their loved ones cope with PTSD and CIS. Fifty-five-year-old Paul MacKenzie, who worked as a first responder with Halifax Regional Police, is the coordinator of HRFE's Firefighters and Family Assistance Program (FFAP).

I have personally known MacKenzie for at least twenty years, beginning when he worked as a police officer and because we also shared a close friend in common, the late former Halifax police officer and city councillor Gary Martin.

Paul MacKenzie graciously welcomes me into his office in Dartmouth, situated near the downtown ferry terminal, where the FFAP program is headquartered. It is located in a 120-year-old, former sea captain's home. It has exposed wood beams, a stone fireplace in a common area, and firefighting memorabilia sprinkled throughout, including photos of *The Patricia*, pre-explosion. There is also a complete living room; the whole effect is calming, relaxing and appealing. It is perfect for the work he conducts.

For our lengthy interview, we sit in a small conference area situated off the living room. It is more like a conversation, actually, because we have known each other for so long. He is a kind and open man who has personally dealt with, and continues to deal with, PTSD.

He says it began while he was working as a police officer and that he still lives with it now, even as he is helping others. "I was involved in CIS management as one who suffered from PTSD and who continues to suffer from PTSD. Once you have it, you have it. It doesn't go away. You learn how to manage it through counselling, and it sometimes involves medication."

I did not know any of this before our meeting.

We begin by discussing some of the roots of PTSD. "The term PTSD probably came more out of Vietnam. We know that people in World War One and World War Two and the Korean War [among many other conflicts] suffer from it. I think the awareness started because of the levels of suicides surrounding those who came back from Vietnam. From there, it's kind of taken off," MacKenzie explains. In World War I those who did not want to go back into the battlefield were considered to be cowards and many were executed. They were, in fact, suffering from PTSD.

MacKenzie also acknowledges the impact of Canadian Senator Roméo Dallaire on the willingness of Canada's military to discuss and deal with PTSD. Senator Dallaire is the highly respected former general and commander of UNAMIR, the peacekeeping force in Rwanda during the 1990s, where mass genocide occurred. Dallaire repeatedly and openly speaks about his own personal battles following his return from Rwanda and other missions.

"He made it okay with the military, because we currently have seen it with those returning from Afghanistan; we are seeing it big time. I've heard him speak several times. And he is one of the first ones to say, 'Hey, it's okay to reach out for help.' Because what he experienced is what so many others had already experienced."

MacKenzie helps define for me what CIS actually is and how it varies from individual to individual. "PTSD occurs from exposure to a critical incident, whatever that critical incident may be perceived to be by that individual. You don't always have to see someone dead. You don't have to be shot, but the fact that you thought you were going to die or be shot, that is a critical incident for that particular individual."

Essentially, if the person perceives his or her life to be in jeopardy, that is legitimately a critical incident for that particular individual. According to the Occupational Health and Safety Division of the United States Department of Labour, it's also witnessing or experiencing tragedy, death, serious illness and threatening situations.

"Often it is a sense of helplessness or hopelessness," explains MacKenzie. "For example, for a police officer or a firefighter, you are trained to go and save. All of a sudden, you can't use that training because of circumstances bigger than all of us. In spite of all the training you have, there's nothing you can do. If you are a firefighter, and another firefighter should die in a line-of-duty death, often those that are with them, and survive, suffer from 'Survivor's Guilt.' They are left asking questions like, 'Why him or her and not me? And, 'How did I get out and he or she did not?' That is considered a critical incident to the surviving first responders."

MacKenzie also offers a checklist, the acute yet most common reactions of first responders, and others, to critical incidents:

1. *Nightmares*
2. *Flashbacks*
3. *No appetite/overeating*
4. *Isolation from others*
5. *Tremendous anxiety about returning to work*
6. *Sleep disturbances*
7. *Irritability with people*
8. *Overly quiet*

MacKenzie elaborates on isolation. "In terms of isolation from others, it is like, 'I just want to stay down the basement and watch TV there. I don't want to answer the phone. I don't want to be around people.'"

The stress about returning to work involves a serious concern about being placed in a similar situation again. "The thing to remember is, it is very much an individual thing. What may be a critical incident to me may not be to you. We just can't paint everybody with the same brush."

MacKenzie says people also internalize how they are feeling. "You can't experience something that is pretty horrific and not feel it. This is all about feelings. I often say when I teach CIS management, 'It's the easiest course you'll have to take. Never, in the next two days, will you have to think. Because it is not about thinking, it's about feeling, especially if you are an experienced person.' It's about feelings associated with those experiences. In my experience, in my professional career, and I've been around a long time, you can't go through something like that and not have it affect you in some form. It may not hit you today or tomorrow, but we don't know what's going to happen down the road."

MacKenzie says a person's reaction can be triggered by something as simple as watching a scene in a television show or film that is similar to their own experience. Or maybe it is something heard or read that happened to a peer or loved one that, all of a sudden, takes the person immediately back in time, reliving what happened to them all over again. "I've heard people say many times, 'I never take my work home.' But, when they come in, they don't talk to anyone for twenty minutes. They have to have their time. Then I'll say, 'Do you take your work home?' Absolutely! You're human and we all do it to varying degrees."

Our discussion makes me stop and think about my father.

Landry never had PTSD or CIS, but, upon reflection, I am certain he had to have been affected by thirty-one years of firefighting in some way. I will say this, my father rarely discussed his work. Many of you, as first responders, or who have loved ones who are, will likely be nodding in agreement. I have new-found appreciation for what my father endured and dealt with after talking to MacKenzie and the other firefighters within these pages.

The FFAP coordinator is also quick to point out that PTSD and CIS affect whole families. MacKenzie says counselling services are available for first responders and their loved ones. He stresses that anonymity is paramount. "Confidentiality is the foundation of this program. Without it, we do not have a program. Because people in this business [first responders] are very mistrustful; whether you're police, fire, paramedics, we tend to be very mistrustful. We're afraid the public's going to think we're not as solid as they thought. None of us want to show that we have any cracks in the armour. People have to feel safe. And it takes a lot of strength to reach out for help. It takes a lot of courage to reach out for help. It does not come naturally to these various subcultures."

MacKenzie's own need for counselling has led him to help others. A thirty-year police veteran, he has also worked with Halifax Regional Police (HRP) on PTSD. He says the police started dealing with it as early as 1985, but it began quietly, among its members. "It was underground. We weren't necessarily supported by management. But, on our time off or days off, we would still provide support to other police officers."

It all began for MacKenzie because of his own struggles and personal demons, "… because of my own alcoholism. And I know that I wasn't the only one to drink. I was the second guy to go through the Employee Assistance Program (EAP) in 1984. We never knew there has been a program since 1977. We never heard of it."

I ask him, "It was never communicated?"

"No," he replies.

During that era it was not communicated with clarity really anywhere.

Drinking to cope was common, according to MacKenzie. It was also a reality in the fire department. My father discussed that fact with me on several occasions. Others I have interviewed also confirm drinking was part of dealing with the anxiety faced on and off the job. It still is an issue today for many people, whether they are first responders or not. It was and is done in many professions, for many reasons.

MacKenzie says, years ago, when CIS education and support had just begun in Halifax, and prior to that time, there was a tremendous amount of peer pressure to return to work after an incident had occurred, far more so than there is today. "Because now, with pre-incident education, we go in and deal with the rookies, before they hit the streets, on what CIS is; here are the signs and the symptoms. We speak about PTSD. We know now to prepare people for what they may experience – not necessarily what they will, but what they may experience. We explain what the spin-offs can be, and how it can not only affect you but the people you love around you. Because it will become a family problem if not looked after."

That has been the case for MacKenzie.

He says his problems began when he was twenty-eight and on the Halifax police force. His drinking escalated. "Drinking was very much part of the culture. For me, it was a number of incidents [that led to CIS]. But there was one, in particular, that sort of brought it to a head – actually two incidents, in a very short period of time."

I have known Paul MacKenzie for many years and have no idea what he is about to reveal. I am sitting at the conference table slightly stunned and baffled that his CIS has never come up before between us. I consider him a personal friend and, because of this, I am slightly embarrassed as we continue with the conversation. I feel like I should have known, should have realized.

It is not the first time MacKenzie has spoken about it. Because of his unwavering commitment to helping others, it is most certainly not his last. This is what happened that led to his own continuing battle with PTSD and CIS.

"By that time, I had been on the job eight years. I just turned twenty when I started. I was involved in a situation where a police officer died. I felt very much responsible. Looking back, I did the best I could. Back then, I got into the what-ifs, could-haves, should-haves. For a lot of years, I lived with that. I remember, exactly, the next day [after the officer's death] absolutely getting hammered and going back down to where the family was. I remember thinking, 'What the hell am I doing here?' He has five children. The five children were there when this all went down. From there, it just got worse. The drinking got worse. I couldn't talk about it and there was other stuff that went down in the department over this particular incident. It wasn't very pleasant and it wasn't handled very well. I was just a mess. November 14th, 1984, I attempted suicide."

The last three words astound me.

MacKenzie stops talking. He tries to compose himself. There is a very long pause between us. It is not awkward. It is extremely powerful. I say nothing.

As his friend, and a long-time journalist, when someone makes a deeply personal admission, you say nothing. You listen. You let the person speak, when he or she feels they are able.

After a brief pause, MacKenzie begins again. "My wife was pregnant at the time. I had a three-year-old, who wasn't that keen on his father, for a lot of reasons. It just started spinning out of control. As a result of that, I knew I needed help. No one knew what to do with me. I didn't know what to do with myself ... After November 14th, I knew I had a big-time problem.

"I remember that day quite well. I knew I had a problem with the booze and nothing was coming together. I mean nothing. I hated work. I hated everybody associated with work. I was having a lot of problems with my family. It wasn't them as much as it was me. I was having nightmares and reliving the whole thing and thinking, 'Why didn't I do this?' At that point I just went home.

I had a .38 and I put it to my head. I remember this like it just happened. I had a misfire."

"What?" I exclaim.

It is the first word I say since MacKenzie starts talking about 1984. I can't help myself. The question just explodes, with concern, from within. I ask it as a friend, not as a journalist.

"Never had a misfire in my career," he responds.

"You had the gun. It is loaded?" I ask, now as the journalist, to clarify and to make sure I hear him correctly.

"Yes. It was my service revolver," he replies.

I ask, "In that moment what did you say to yourself?"

"'Fuck, I can't even do this right,'" MacKenzie answers.

The frank and poignant response is said entirely without dark humour and with great sorrow. It is truly how low MacKenzie felt at that desperate moment.

"I didn't try again. I knew I needed help," he adds softly.

Twenty-nine years later, as he relives his private hell with me, I respond this way: "Oh, my God! I have to hug you."

We stop the conversation.

I get up from the table and go over and hug my old friend. We both break down. It is a powerful and moving moment I will never forget. I have conducted thousands of interviews and I can count the number of times I have broken down on the job on one hand.

"I couldn't imagine if it had gone off," I manage to say, through tears.

"I think I've kind of found out why it didn't," he quietly concludes.

MacKenzie discovered his true calling the hard, hard way. He has become a living testament for others that there is life, a rich and rewarding life, while living with PTSD and CIS.

"We now have best practices. We will touch base with firemen after difficult calls – people like me, and other firefighters, who have been trained to provide support, and who have been well educated in CIS and PTSD. Early intervention plays a part. If nothing else, you can plant the seed that people care."

He says the help must begin immediately following a tough call. "Often we will either go to the scene or meet them after at the station. What they do now is, they take a station out of service until we can do our thing and provide them with support. We talk about it before they go home; start to do a defusing, acknowledging what they have experienced. We talk about the real facts, not as how we perceived them to have been – 'This is what happened.' We do all that before they get off shift. Then, within the next day or two, we will organize a formal debriefing, if it is required. If you need professional help, we set that up. I make the referrals. We pay for it. There is no cost to the firefighter. We look after all of that."

"What about families of firefighters? What if they need help?" I ask.

"You're part of the family."

The "you're" MacKenzie is referring to are all family members, including me. It is the second time during my research that a first responder has told me, point blank, that by being the child of a firefighter, I am considered part of an extended "family." It has helped, in a small but meaningful way, to heal my own loss.

My father retired in the late 1980s. By that time, MacKenzie says the fire department had a system to deal solely with CIS management. "It wasn't necessarily working well. They had a really great group of firefighters that were doing this in a program and doing the best they could. That was 1985 to 1988."

When MacKenzie was still working with Halifax police, he also dealt with a lot of firefighters. He describes them as hardcore alcoholics and those with suicidal behaviours. "A lot of people involved in the program and helping were in recovery. We all had our own struggles of one sort or another. That kind of made it easier for people to reach out. Often people say, 'Well, you don't know how it is!' I'll say, 'I don't know how you're feeling, but I think I got a pretty good handle on it, what it was like, or what it is like.'" MacKenzie says when the first responders finally realize he's gone through similar issues, then they open up.

History reveals and MacKenzie explains that Halifax police were initially farther ahead in dealing with PTSD than the fire department. He says, early on, the police received open support from former police chief Vince MacDonald, who started in that post in 1990. "We were fortunate we had the support. Now, the first six years, we worked underground. But when Vince became chief, he said, 'We've got to bring this above ground.' I was fortunate I was given the opportunity to do a pilot project for one year under Chief MacDonald. That was in 1993. And from there, we never looked back, because it became a permanent thing."

A decade ago, MacKenzie was contracted to provide support services to volunteer firefighters in the Halifax Regional Municipality. That was deemed so successful, he says, he was also asked to help develop the current program for professional Halifax firefighters. "Prior to 2003, they [fire] were only doing CIS management. This program is now all encompassing. I also look after veterans [retired firefighters]. We're starting a Veterans' Association up now. These crests just came out."

MacKenzie shows me a crisp, new crest that looks like a badge. They have been created for retired HRM firefighters. "We also provide support to the veterans. The biggest thing is to try and communicate that. One of the traditions we've also just started, in 2012, is that the junior person in the fire station presents the retiree or veteran firefighter with the crest. The informal ceremony takes place in whatever station they retire from."

At this point, MacKenzie hands me a crest. He presents it to me on my father's behalf. I gratefully accept it and tell him I will give it to my mother as a surprise gift.

I decide to give it to Mom on Mother's Day. I placed it inside a beautiful, shiny red keepsake box, the traditional colour of fire engines. It is accompanied by the following write-up. I created it for my mother to explain the presentation and the exact moment:

"Veterans Crest History and Presentation: Halifax Regional Fire & Emergency started presenting these crests to retired Halifax firefighters in 2012. It is also significant that they are calling the firefighters 'veterans.' Paul MacKenzie, coordinator of the Halifax

Regional Fire & Emergency's Firefighters & Family Assistance Program (FFAP) says there wasn't a dry eye in the house when the first presentation was made (to the first retiree in a fire station). The crest you are looking at was presented to Basil Landry's daughter, Janice, on May 7, 2012, by Mr. MacKenzie in a private meeting/interview when she was working on her manuscript to honour her late father and all first responders. There wasn't a dry eye in the house at that ceremony either."

My mother loves the crest and the backstory of how we have come to own one. The red box now sits open on a bookcase in her living room, next to a picture of my father sitting in the front seat of a fire engine dating from the 1950s, when he first began his career. Halifax Regional Fire & Emergency has certainly come a long way in dealing with PTSD and CIS since my father's era.

As of April 30, 2013, MacKenzie says, there are now four chaplains, twenty-two career firefighters who volunteer as FFAP referral agents, as well as volunteer firefighters who also give of their time as agents. "We have two people who are on call 24/7."

MacKenzie is the only paid FFAP member. Everyone else is a volunteer.

"This is a very busy program," he says. "We do a lot of referrals. Since I came in 2003, we have made more than 836 referrals, men and women. That's not counting the people we've met over coffee or that drop in here. These are the people who actually went for referrals. We also want to get the message out that veterans can also use our program and that veterans have gone though the program. They are still struggling."

MacKenzie has been working with career HRM firefighters since 2007. Despite the pain from his past, he considers himself lucky; he does not consider his work with FFAP a job. He maintains, if it ever becomes a burden he will be forced to leave his post.

Twenty-nine years after his failed suicide attempt with his own gun, Paul MacKenzie sums up that November day and what has followed, with conviction. "I have a better understanding of how I got there. I didn't at the time. I do today. I always tell people it [FFAP] wasn't my career path. It just worked out this way. But, like everything else, I always have to do it the hard way. If you'd asked me, 'Would I do it all over again?' No. But those experiences have led me to where I am. I am glad I went through them. Would I care to go through them again? No."

Chapter 8

Into the Darkness

> *So nigh is grandeur to our dust,*
> *So near is God to man,*
> *When Duty whispers low, Thou must,*
> *The youth replies, I can.*
> ...
> *Peril around, all else appalling,*
> *Cannon in front and leaden rain*
> *Him duty through the clarion calling*
> *To the van called not in vain.*
> — *Ralph Waldo Emerson*

This quotation from Emerson, one of my favourite writers, is from his poem "Voluntaries" and also appears in his famous work *Society and Solitude*. It is his preamble to chapter ten, which he dedicates to the topic of courage. It seems appropriate at the start of this chapter, which also describes acts of courage and selflessness in the face of death and despair, and the repercussions from both.

I have noticed a common thread among my discussions with first responders for this book. Most, if not all of them, are able to recollect details, even the tiniest ones, from fires they fought many years ago.

This chapter provides very frank, first-hand accounts about some of the toughest calls these men have dealt with and the effects, both mental and physical, of working as a first responder. I would like to caution readers that some stories may be upsetting. A realistic discussion about this part of the job must take place because it is the grim reality of this work. Furthermore, it was not discussed for far too long, which has led to many volunteer and professional first responders not receiving the help or counselling they need.

I am beginning with Rob Brown because it is significant that he discusses the impact of the 3390 Federal Avenue fire despite the fact a rescue took place and no one was injured.

Rob Brown

On what happened at 3390:

"This is my first search. I'm the only guy in there [he thought] at that time when someone tells you there's a baby in there. You know the place is charged with smoke. Therefore, there's no oxygen to breathe, especially with an infant. I'm thinking, 'I've got to find that child, the child is not going to survive, and I'm the guy in there and I can't find it, and I'm inadequate because I didn't do what I *should have* done.'"

Brown talks about the what-ifs, should-haves, could-haves that we all ponder at one point or another.

"But you did, though." I mean that he did, in fact, make it into the boy's bedroom to find both Landry and McKenzie.

"I did, but *what if* I got in there and the child was deceased?"

As soon as the words are out of his mouth, I can't help thinking if that had happened, Brown would most likely have found the child and Landry both deceased, lying together. He was the only one inside the fire and the room. The *what-ifs* also run through my mind.

We all know that a tragedy did not happen. So what did happen when the firefighters returned back to the station that day in October 1978?

"When you go back to the station and lives are saved and everybody does their job, it's such a bonding experience," says Brown. "It would be the polar opposite if things went wrong. It would be such a bad, bad feeling. People who don't do that job, you can't explain thoroughly enough the bonding experience that goes on. They're your family in that station."

Ron Horrocks

Former fire chief Ron Horrocks addresses the issue of PTSD during the primary part of his tenure. "First of all, we didn't even know what that was. We didn't have a clue there was such a thing. The ones that I knew, myself even, the way we dealt with it was to bury it, and most often with false bravado, even joking about stuff. That's just the way it was. But that's not to say that a price wasn't paid, down the road."

Horrocks started his lengthy firefighting career in the 1950s as a hoseman in Québec. His first story dates back to his inaugural year on the job.

"I can give you a good instance. It didn't happen in Halifax. It happened in Montreal. I was on the Westmount Fire Department. It was 1956. We had a fire in an apartment building and the roof was pretty well involved, but it hadn't broken through. The building had been reroofed. The fire was two layers down, a lot of heat. The firefighters put up a ladder and they went up the roof. This one fellow was just going to open the roof to ventilate it and he fell through. He fell in!

"The guys who were with him laid on their bellies and felt around for him because smoke was now coming out of that hole. They got him and pulled him out. He said he was fine. I spoke to him about this. He worked the fire the whole time, and it was all night. He said when he went home, 'I was fine. I was okay. I just worked and it was behind me. I sat down with my wife later and picked up a cup of tea and the teacup started to shake.' His hand couldn't hold it steady, the aftershock. He came close."

The former chief remembers his own aftershock and upset over the loss of a child. "I remember coming home and sitting there with Marina [his wife] and I burst into tears. And it was over a kid, you know? It was a youngster who had died. You may not think about it at the time, but sometime afterward, it comes out. It's going to come out. You have to purge it."

The one call Horrocks says he will never forget also happened in Québec, also early in his career. Coincidentally, many of the fire-fighters have chosen to discuss their first calls where loss of life has been involved. "I attended at an explosion in Québec. It was in the late 1950s, early 1960s. It was a three-storey apartment building. It was what they call a pancake collapse. People were trapped between the layers of the floors which were now in the basement. I don't remember how many people died, but we were pulling people out, lifting stoves off people, fridges, et cetera.

"There were some we couldn't get out. That, I remember. I re-call two of us, this other fella, he was as strong as an ox and he couldn't lift a hot water tank off somebody. They were upside down and we had a hold of them by their legs. They died right there in our hands. You never forget that stuff. But don't ever lose sight of the fact that there are plenty of other occupations where people do the same thing."

I have been able to find information about two Québec apart-ment fires during Horrocks' firefighting tenure in Montreal. They are both referenced in a paper published online, written by Robert L. Jones, entitled *Canadian Disasters – An Historical Survey*.

The first is a fire at Olfields Apartments that happened on No-vember 7, 1958. Eighteen people died. The online resource indicates weather was not a factor but no other details are discussed. The sec-ond one involved an explosion in LaSalle, a suburb of Montreal, in 1965. Twenty-eight people lost their lives. According to Wikipedia, "LaSalle Heights Disaster occurred in the early morning of March 1, 1965, in the city of LaSalle, Quebec, when a gas line explosion destroyed a number of low-cost housing units. In all, 28 people lost their lives. 39 were injured and 200 were left homeless. Most of the casualties were women and children because many men had left for work. The casualties might have been higher had it not been the first of the month when many men left earlier than usual to pay their monthly rent at the rental office."

The Québec apartment collapse is still very much at the top of his mind for Horrocks during our interview, nearly half a century later. He concludes, "In those days, if you discussed stuff like that [PTSD] it was seen as a weakness. You didn't do it."

My own father's lack of discussion about his work is making far more sense after talking with his former colleagues.

Don Swan

Don Swan began his firefighting career with the Halifax Fire Department in 1955. Despite the passage of time, Swan can also re-count many stories and fire calls in minute detail. "There's always one where you can feel 'old death' at your back. I didn't come out of the fire. They got me out. I was deputy chief. Ruth [his wife] was with me in the car and we were going out somewhere and I got the call on the radio that there was a major fire at a multiple unit, old building down the south end. It was in the early 1980s. I can remember what we were trying to do.

"Jack Pritchard was the captain. I was the deputy chief. There were two other firefighters. We were trying to get through the at-tached buildings because someone told us there was a woman trapped at the other end of the corridor. You had to go up three steps, the way the building was divided. We got up the three steps, but, of course, we were going up into the blacker, heavier smoke. We made three attempts. We got up, stopped, and backed down again. I kept saying, 'The woman!' Pritchard and I were the two of-ficers. He said, 'We've got to get out of here!' So they turned and went.

"I stupidly said, 'I think I can get to her.' So I kept going and didn't leave with them. I went up the steps and tried to crawl. I heard church bells ring, stuff like that. You hear a sound like chimes. It's lack of oxygen. I was hearing the lack of oxygen in my head. It's a saying [among firefighters]: 'You hear bells ringing to get out.'"

Swan says he does not remember his comrades coming back into the inferno to save him. "Jack Pritchard realized I wasn't behind him. He came back in … I remember being outside on my back. I think Mo Cannon was giving me mouth-to-mouth." The 81-year-old, as of 2013, says he will never forget the sound of the bells or that his friends risked their lives to save his.

Swan also does not forget the cases when the outcome was tragic. "I had children, babies die, stuff like that. You try not to think about it. You come home and you look at your own kids and you get cold chills. I've been retired since 1988 and I still have flashbacks and nightmares. A lot of it was caused by angst."

The angst Swan refers to is relentless and overpowering for many. Swan says he personally knows firefighters who drank heavily trying to escape it. None of this should seem outlandish or surprising, given the admitted lack of understanding and treatment for PTSD and CIS years ago, and because of what all first responders still routinely face today.

Swan also opens up about a friend of his who fought a fire in the 1980s that haunted the now deceased firefighter for the rest of his life. Out of respect for the late firefighter and his family, I have chosen not to reveal his name. It serves no purpose. What does have deeper meaning is that we can reflect about what happened to him and try to reach out to others facing similar circumstances. For the sake of storytelling, I am changing his name to John Doe. Swan recounts his story. "I was home here and it was Saturday morning. The phone rang."

As a senior officer later in his career, Swan was always informed of fires while off-duty. This one was considered a Code 5-3, which the retired chief and deputy chief explains. "A 5-1 was a standard fire that you could handle. A 5-2 was a working fire and 'We may call for help.' A 5-3 was 'Help!' Harold Young hollered out on the radio, 'They're hanging out the windows!'"

A horrific scene was unfolding at a major fire in an apartment structure. "We lost five in that – a woman, her friend and a little child, and two men in another unit. We didn't lose any firemen that day, but I know one guy, John Doe, who died that day because he never really got over that he couldn't get [to] the guy on the top floor [in time]. Took the heart and soul right out of him. He couldn't reach the guy [in time]. … He has his air pack on and it was terrible in there. … He got in and was searching for the guy and could not find him. He ran out of air and he got to the window. He slid the window open and pulled his mask off and hung his head out. I called to another fireman to get a ladder and I was going up. And he said, '[You're] higher rank, Chief.' And he went up. The other firefighter [Doe] came [out of the window] head first." Swan explained that Doe's face was discoloured from oxygen deprivation. The fireman on the ladder stayed with Doe as they both came down.

The memory of the tragic fire became too much for Doe, according to his long-time friend. "He used to come and visit with me here and … talk about not getting the guy [in time]." Doe was haunted by his experiences during his career. He did not get help for his PTSD and suffered to the end of his life, especially over that particular traumatic event, tortured by the what-ifs, could-haves, should-haves.

Joe Young

Joe Young's two stories date back to the 1970s. A former fire department driver who knows every street and back alleyway in Halifax, Young is very detailed when he describes the location of the two calls.

"One of the big ones was Speedy Propane. Jack Pritchard and I were on the rescue [unit] and we had a call for a boxcar fire behind a fence on Dutch Village Road [in the city's west end] where Joe Howe Drive and Dutch Village Road meet.

"I'm driving up to it and all of a sudden I said, 'Jumping Jesus, that's Speedy Propane!' I whipped the wheel around to try and go the other way. But because of the railway tracks, I couldn't make the cut. I parked the pumper sideways and that's where it stayed. We didn't want to park any closer. There was a loading dock where they used to bring the propane tanks in. It wasn't a boxcar, it was a rail car. But it was a propane rail car on fire.

"All of a sudden, the cylinders started to blow! They'd go up in the air and burn and pop like a giant firecracker. But those cylinders had to come down. They were 300-gallon cylinders, big round ones! I could see a couple of them come down, but then I lost sight of them in the fire. They're coming down, hitting the railway tracks, at two in the morning, and making a large booming sound. You'd hear a whistle and you'd know one was going up in the air, but you'd be thinking, 'Where are they coming down at?'"

Miraculously, despite the dangerous debris flying around their heads, no firefighter was badly hurt. "Jack Pritchard and someone else got a Medal of Merit for that one. I didn't think I'd ever come home. I said my prayers that night, I'll tell you."

It did come very close for Young. "One of the tanks went under a trailer with a wooden floor. I went up onto it [the trailer] to turn a [feed] line on the fire. One of the cylinders blew and the concussion [from the explosion] just knocked me right over, off the top of the trailer. I got up and said, 'Wow!' and I tried it again, because the fire was still raging. The other firefighters were all busy running a line from Windsor Street down through a cemetery. There was an open grave and one of the firefighters fell in and got hurt."

Despite the close call that almost cost Young his life, it is another emergency, an accident during 1970-1971, that Young says is one of two cases that still cause him to experience flashbacks.

"We had just got the Jaws of Life and there was a car accident on the Bicentennial Highway [Bi-Hi, Highway 102]. It was December 23rd. The Bi-Hi was just being built. The other side was being twinned. We went out with a rescue [truck] to cover the Rockingham fire station."

Firefighters were often called to cover off stations if the members from that company were already out on another call. "I ran across the street and I got a bunch of doughnuts. And when I was coming back they said, 'Joe, we got to get out there!'"

His shift-mates told Young en route that he may have to use the Jaws of Life to extricate a driver from his wrecked car. Young says he had two training sessions with the newly received life-saving equipment.

"We went out there and it was a pretty gruesome scene. One of the ambulances was there and they pronounced him dead; the guy was still sitting in the car. He hit a tractor trailer. I can describe it, but you don't need that. Anyway, the V.G. [an ambulance from the Victoria General Hospital] showed up. The V.G. got a pulse. They said, 'We've got to get him out.' We got the Jaws out and cut the roof off. He left for the hospital.

"Anyway, I had a box of doughnuts and we went back and we sat on the tailgate [of the fire truck]. This is one of the things with critical incident stress, this is where I am going with this story. I opened it [the box] up and said, 'Look at this, jelly doughnuts, we just saw that.' Two of them took off the side of the pumper [to vomit]. I could hear them over there. I shouldn't have said it. That was my way of coping with the moment because I was standing there, cutting with the Jaws, and I had to look at this guy. The whole side of his face was exposed."

Ron Horrocks also says dark humour was a coping mechanism for some. My father told me the same thing. As difficult as it may be for some of you to admit it or accept this fact, it is part of the very tough reality first responders face when they are forced to witness horrific accidents and emergencies.

The accident and the man's death affected the entire responding crew. Young says they later discovered more about who the driver was. "He was twenty-one or twenty-two. He landed a job in Halifax and was from Cape Breton on his way home for Christmas. The car was chock full of presents. He was taking all of this home for his family." Young paused. "I think about that ...," he says softly.

The veteran firefighter sometimes experiences flashbacks stemming from another call to the Bi-Hi where, in this case, firefighters did receive some counselling afterwards.

"The guy came on the highway at Lacewood Drive and came out onto the highway the wrong way. He hit two cars and three people died. One of them was in the car and we couldn't get him out. We could hear him hollering. Father Lloyd [O'Neil] came in that night to talk to us. We all went in uniform, in a fire department vehicle, to the young man's funeral. Father Lloyd also had a couple of other sessions with us."

Young rose through the ranks and became a platoon chief before retiring. It was during this latter part of his tenure as a senior officer that a horrific tragedy occurred at Peggy's Cove.

In 1998, Swissair Flight 111 crashed into waters about eight kilometres from the famous landmark. On the night of September 2, all 229 passengers and crew died when the plane plunged into the Atlantic Ocean. A lengthy investigation found a fire had filled the cockpit and plane with smoke. Young says he arrived on the catastrophic scene on day two. He starts the story by explaining what he had heard the night of the crash.

"I went outside to have a smoke and it was 11:25 p.m. I could hear the plane. I thought, 'That's strange, it's circling.' I heard it. It was some low. All of a sudden, I didn't hear any more; never thought about it. The helicopters started coming over [his house] from CFB Shearwater. I could see them coming with their big lights on. Two went by. I wondered why they'd be out here. So I turned the news on. They said a plane went down at Peggy's. I said, 'Oh my God, I bet that's the noise I heard. That's the plane!'"

At the time, Young says he was the acting Emergencies Measures Organization (EMO) co-ordinator for HRM.

"I was with HazMat [hazardous materials]. I was platoon chief but HazMat was my speciality at that time. So we had to put a team out to help with recovery. We also had to put a team out there for firefighting because that village went from eighty-five residents to 1,800 to 2,400 people at times. So you had to make sure you had fire protection in the village.

"They asked us to provide a crew for the barge. The barge was out there in the water with the claw bringing material up and putting it on the deck. They wanted a team out there to help with decontamination. We knew it was going to be pretty gruesome. That was one of my hardest jobs. I had to hand-pick four crews of five guys each to go out there on twelve-hour shifts, and to work that. I went through the list saying, 'This guy can handle it,' thinking I know these guys. So we called them all in and interviewed them all. Some we eliminated; anyone with small kids we didn't take, or … we didn't want to take those ready to retire in case the stress got too much for them. There weren't many nights I didn't come home and think, 'God, I hope those guys are okay.'"

The responsibility and stress worrying about his crew weighed heavily on Young. He says he knew he could not face the barge himself. "I didn't go out on the barge, by choice. I thought, 'If I go out there and something happens to me [in terms of a reaction], I'm going to let these guys down.' Anytime those kinds of incidents happen there is a percentage they figure they will lose to CIS."

Young says it was FFAP's Paul MacKenzie who did the debriefing with the first responders Young had helped choose to work on the recovery barge. "He did a great job. It was a mandatory thing for our crews. Before we would pick them, we would say, 'Now this is the mandatory part of it. You have to do this after the fact.' We also had two meetings right after. The next year we did a follow-up and also a follow-up in the third year."

I was working at CTV Atlantic when the Swissair crash occurred. Initially, media organizations from around the globe swamped our Halifax newsroom with calls. I fielded many. Then news crews started arriving. I worked the first few days in CTV's Halifax newsroom helping to coordinate the calls and coverage.

It was difficult for our staff who went to the scene because of what they saw in the first few hours and during the many ensuing days. I viewed some of their images through unedited, raw footage they brought back to the station that was never aired; parts of it were too graphic for public viewing.

I remember being seated in an editing suite reviewing some of the videotape and seeing images I still wish I had not; they are forever burned into my memory. News organizations also have employees who suffer from PTSD and CIS stemming from stories they cover. I have never required counselling services myself, but there are certainly many stories and sights I will never forget that still affect me today.

For example, I covered the crime beat for many years and eventually decided I needed to stop because I found it very taxing and emotionally draining to always be discussing murders, stabbings, shootings, sexual assaults, among other horrific incidents. The one blessing during that challenging period in my career is the many people, from all walks of life, who spoke to me, some anonymously, to try to help a particular person, family, or the larger community in some way in the face of tragedy. Those eye-opening years earlier in my career were demanding, but they have helped educate and shape me.

As part of my journalism work during the aftermath of the crash at Peggy's Cove, I had an opportunity to interview, several times, a superb member of the counselling staff with the Canadian military at CFB Halifax (Stadacona) about the effects of PTSD and CIS on its members. The military was also called in to assist in the aftermath. I have never forgotten the woman I interviewed telling me, on-camera, that she was aware of people who were unable to return to their posts because of the impact of what they had dealt with over the course of the Swissair recovery and investigation.

John Fitzgerald

Sixty-seven-year-old John "Fitzy" Fitzgerald started with the Halifax Fire Department in November 1968. Fitzgerald knew my father well and, like Young, also drove him and other officers to and from fire scenes. Fitzgerald, his daughter and their families attend the same church as my family and where my father's funeral was held. As a result, Fitzgerald has become a friend and is the first firefighter I interviewed for this book, in January of 2012. He is the person who first suggested I approach firefighting historian Don Snider, and this work took off from my initial interviews with both of them.

Fitzgerald is funny and outgoing. However, his voice quiets when we sit together in his kitchen over a cup of tea, while he discusses firefighting, my father and one old call, in particular.

While Fitzgerald was a junior hoseman, he went to this emergency his first year on the job. "I was twenty-four years old. I was out at a station in Spryfield, near McIntosh Run. A little boy, six years old, went through the ice. I did everything I could possibly do. It was in February. I wasn't even on the job six months. I went out spread-eagled across the ice so I wouldn't fall through because his school bag was there. I got the school bag, but he wasn't on the end of it. We had to get a diver to go down and bring him up. There was nowhere to go to talk about it."

More than forty years later, Fitzgerald still thinks about the little boy.

Chris Camp

Forty-four-year-old Chris Camp started working as a firefighter in 1990 at the Bayers Road station, the base for the responding crew for Federal Avenue back in 1978.

"It's a rewarding job and a crappy job at the same time," he says. You see some things that are unpleasant. I had to hold someone who was shot in the head, ... car accidents, people who are dead at the scene and you try and do the best you can. Our guys, a few years ago, spent hours cutting somebody out of a car so that the family could have some closure. Because when a car is mangled around a person, it is not a pleasant thing to get them out. We've had some guys who've done things they don't have to do, but they want to show respect for the person and for the family. Sometimes you do your best, and it works out and sometimes it doesn't. It's just fate."

Camp recalls one frightening scene when his colleague came close to dying. It was about two o'clock in the morning. He is not sure of the exact year but says it happened pre-amalgamation.

"I was there as part of a crew when another one of our members went through a roof in the middle of the O'Brien's Pharmacy, across from Howe Hall at Dalhousie University. They think it was an old chimney that had been finished over. It was like a trap door that burned away. He stepped on it and went right through. Just through luck, he landed between the aisles on the floor. He said that he ran for the only thing that wasn't orange, which ended up being the door. He was fortunate that he was in an aisle, and the aisle was pointing towards the door. He didn't come out right away. One guy tried to punch a hole in the wall to get him. He had fallen about eight to ten feet and was disoriented and shaken up. He doesn't talk a lot about it. It was a close call. The flames shot up, where he went through, next to the guy beside him. They looked over and he's gone. They stuffed a feed line down to try and cool down wherever he was. Sixty seconds later he was out the door. He was not injured."

It's another frightening sixty-second story with a positive ending.

Bob Whorrall

Bob Whorrall, now sixty-nine, was involved in one of the types of cases that Camp has just described, ones where first responders push themselves to the limit for the benefit of the families of the deceased.

The case Whorrall describes to me happened in the 1970s, when he was still working out of the Bayers Road fire station. Whorrall was called to an accident on the Bedford Highway near Fisherman's Market. A milk truck had collided with a Volkswagen. Firefighters used a cutting torch to try to free the two decapitated passengers. Whorrall was the person who had to place asbestos blankets, through the windows, over the man and woman as firefighters worked at freeing them.

Despite the fact the accident scene played out more than forty years ago, the veteran firefighter concludes, "I can remember every horrific thing that ever happened to me."

Doug Findlay

The 80-year-old firefighter reflects back to the year 1955. The location of the fire was the former Africville, which was located near the present-day A. Murray Mackay Bridge, along the shores of the Bedford Basin in an area now called Seaview Memorial Park. The veteran of thirty-two years was just twenty-two years old that day.

"There was no water up there. We had to run long feed lines from Barrington Street. It was a small house that was completely engulfed."

Almost sixty years later, it is the odour emanating from the home that Findlay cannot forget: "… the smell of somebody … little fella burned to death … what you're thinking of is your own youngster."

Findlay had a one-year-old son at home.

Gerry Condon

Gerry Condon was twenty-two when he joined the Halifax Fire Department in 1969, during the city's first amalgamation. Condon is well known among his peers for organizing many of the Ladder-a-thons which firefighters have routinely conducted in Halifax's Grand Parade, a military square in the city's downtown core. The events are fundraisers for the various charitable causes the firefighters support.

Condon has also chosen to discuss another call involving a child. "I was on the rescue [truck] a lot, and we had a call one night from West Street [fire station] over to Clifton Street. And it was a baby – a plastic bag above the crib. The baby suffocated. I did know mouth-to-mouth because I had learned it in the Armed Forces. When we came back [to the station] we all sat on the tail of the rescue. We were all smoking and crying."

There is another call, one Condon says he has consciously worked at forgetting. "One of my first calls, when I worked out of Spryfield, was an automobile accident on the Purcell's Cove Road near Oceanview [Drive]. It was a car in the ditch and it was turned and facing back to Halifax. The car rolled over and his [the driver's] head went outside [of the car window]. It crushed his head. Dave Starrett and I got the car lifted up. When you see something like that – it's there – I can still see the car. But the head, the vision of that, is gone. Eventually, you work on a whole bunch of things to get them [the sights] gone."

Bernie Harvey

The detail with which Bernie Harvey tells a story from the 1970s is astounding.

"I was a lieutenant at the Dunsworth's Pharmacy, Quinpool Road, fire. It was the early '70s. The fire started in the basement. We took a line down. I was at West Street and it was a Sunday morning, in August, around nine in the morning. We had just come on duty. Up on the first floor of the pharmacy, they had one of those old-fashioned showcases made with glass. It was ten to twelve feet long. And what happened is, the floor weakened. We were down in the basement and the showcase came in. It broke through the first floor of the drugstore into the basement. You're always trained to follow the hose line, to crawl it. So I had to crawl it. I lost my helmet and everything. I had very little room under the showcase to crawl under it. And this is in smoky conditions and everything. There was about a foot clearance. I could see vaguely. It became lighter as the smoke lessened. There is no time to think about anything. Just to get out!"

Harvey also describes another call that came into the Morris Street fire station, in the late 1950s. This one did not have a positive outcome.

"It was 1958-1959. We had a fire down on Kent Street ... It was two to three in the morning. A girl lost her life in there. We saw her in the top window. They put a ladder up. We went up the stairway. Everything happened so quickly. She was in the bay window. The fire was in that room where she was. We got to the top of the landing and there was a backdraft and it blew us back down over the stairs. We rolled right up with the hose on the next floor landing. She died in the fire. She was nineteen."

Harvey was badly injured in a 1957 fire, the physical effects of which he still deals with. "I got my eyes burned so badly at the College Street fire. There was a rehearsal at the old College Street Theatre. It was the coldest winter night. I'll never forget it. When we got there, the whole back of the building was going. It was all two-and-a-half-inch lines then. Not one-and-a-half-inch. I got wet right away. We ran lines through the front. All you could see was a glow coming up from the basement. It burned right through the whole floor right in front of us. The whole floor let go. If we had have been inside another ten feet, there would have been four or five of us gone down into the fire.

"We left and backed out. It was a three-storey, wooden building, with a bell tower on it. It was so fast. When I got onto the front steps, it [the fire] was up into the bell tower. They told me to take an axe and get the screens off the basement windows. I did that. By that time, the fire is up through the roof. There were two aerials there: one from Morris Street and one from Bedford Row. All the smoke and cinders were coming my way. I had no breathing apparatus and all the wind was blowing back towards me, on top of the aerial.

"When I finally got relieved on there, my clothes were frozen on me. It was cold. They had to put hot water on my buckles back at the station to get my fire coat off. I went home. My eyes were burned. I still have trouble with my eyes and put eye drops in every day. I have dry eyes. I had to put drops in my eyes for three days afterwards. I could see blurry. I had to go to bed for three days."

There is a lingering physical toll from that fire and a lingering psychological one from a call, years later, when he was platoon chief.

"There was a fire at Belle Aire Terrace where we lost a baby and a mother. Psychologically it bothered me. I can say that. Small little baby, only six months old. I was stationed at Central [HQ] and we responded. They got up on the second floor and they found the baby and the mother, and the father was over them. He was burned. He ran to the back of the building and they went to the front. They died towards the front on the second floor, facing North Street. They put a twenty-four-foot ladder up. They passed the baby to a firefighter and he passed it to me. I remember laying the baby down and the paramedics were there. The lady came down the ladder like a rag doll. She died of her injuries. She was out of it when they brought her down on the ladder – from the smoke. We put the resuscitator on her. There was no life. The baby died in hospital two days later. It bothered me."

Harvey's next account underlines, like my father's 1978 rescue, that one minute can, and does, make the difference between life and death.

"I was working out of Morris Street. We went to Clyde and Birmingham streets. A man is hanging out the window. A twenty-four-foot ladder will not reach three stories, but we had to try something. The aerial was still on its way. We tried to put the twenty-four up and tried to hold him in until we could get the aerial, or get up the back of the building. He fell out. He got top-heavy leaning out trying to get fresh air. He was overcome. He kept inching out for fresh air. As soon as someone gets smoke condition and hits the fresh air, they become overcome. It was a mattress fire – drinking, you know, cigarette, you know. I never smoked in my life. Thank God. His head hit the concrete and it bounced in front of me and tore it right open. The fall killed him. I can still see his head. He came down about two to three feet from my rubber boots. Those are things you always remember. I remember his name and how old he was."

Harvey does not share those details and I know not to ask.

In fact, when I discuss all of the calls in this chapter with the firefighters quoted, I did not ask many questions. I simply listened.

"I never discuss it [these memories] with anybody," Harvey quietly says of the Clyde and Birmingham street blaze. "If we had got the call one minute earlier, we would have had a bigger ladder to get that guy out of there."

It is another example of the would-haves, could-haves, what-ifs FFAP coordinator Paul MacKenzie explained. Unfortunately, we hear and read about tragedies far too frequently, but society is becoming more willing and able to deal with the long-term impact on all those involved. As Paul MacKenzie says, work is being done to help prepare first responders prior to these types of calls occurring. The attitude is becoming more preventative rather than reactive.

The late Ron "Brownie" Brown, through his daughters Heather Brown Harroun and Beth Mader

Ron "Brownie" Brown joined the fire department in 1959 and retired in 1989. His daughter Heather Brown Harroun says her father died September 23, 2000. He succumbed to small cell cancer, caused by asbestos.

The type of cancer Brown had, Harroun says, is considered a "firefighter cancer." After her father's passing, her late mother, Anna, diligently lobbied and worked to have her husband's case and illness officially linked with his work as a firefighter.

According to Harroun, "A number of specific cancers have been recognized as firefighter cancers under Workers Compensation, and, as such, have been specifically identified as ones that there is a cause and effect between the job and the developments of these cancers ... Mom went the extra mile after Dad died to have a ruling on his case, as she felt very strongly that it should be recognized that he, like so many other firefighters, had given his life in service ... while it may not be a traditional 'line-of-duty' death, his service had ultimately taken his life. He would have been proud of her, but he would have never made a big deal of it ... he was simply doing his job. Dad was very humble and private but he was a great man ... I miss him so," says Harroun.

Brown and his family lived in the same general area of Halifax that my family did when the 1978 rescue took place. Harroun and I went to school together and also attended the same Christmas parties that were held for the children of firefighters.

Harroun, her sister Beth Mader, and their late mother believe there is one key fire where Brown came into contact with asbestos, which eventually led to the firefighter's death from cancer. "I consider Dad and all of those who serve [as] heroes, simply because they step up and say, 'I'll be here for you and on my watch, I'll do my very best for you to not let anything hurt you.' With that, they

not only put their lives on the line, but they see such terrible tragedy ... I take great comfort and solace in the many lives he would
have touched over his thirty years of service ... how privileged I
know he felt to have served ... as hard as it was to lose Dad to a
firefighter cancer ... looking back now, we realize we were living on
borrowed time with him, as he was almost killed when I was two.
A ceiling collapsed and he almost didn't make it out when a washer/
dryer landed on him. His buddies pulled him out. We figure given
how long he was trapped and the debris ... there is probably a good
chance that was when he took in the asbestos," explains Harroun.

She still vividly remembers the calls and cases that impacted
her, her father and the Brown family: "To this day, the pumper or
the aerial doesn't pass me that I don't tear up and wish the guys
Godspeed. This feeling, perhaps, was never as vivid as on 9/11. I
had got home from dropping Kate [her daughter] off at school, and
Maddie [her younger daughter] was a baby and I was on maternity
leave. I had just turned on the TV and watched as the second plane
flew into the tower. I remember my very first thought was: 'Oh my
God, so many firefighters are going to die today!' It was almost a
year that my dad had passed away and those emotions were still very
raw.

"I remember Mom saying, on the night of 9/11, that she was
glad Dad was spared seeing that, as he would have been crushed at
the loss of so many brothers. I think men and women who are in
service are truly called to it; Dad never really did discuss the close
calls or bring too much home with him ... and although he didn't
talk a lot about the job, we could normally tell when he was having
a hard time with something he saw or experienced."

She recalls one time when Brown did have something to say,
making his point loud and clear. "Funny story... Stephen Bowser
[a friend] picked me up on his motorcycle one summer to go for
a swim at the Policemen's Club. When Stephen dropped me home
later that afternoon and Dad saw me on the back of a motorcycle,
he flipped. He went into a long tirade with examples of the accidents he had seen with motorcycles on the job ... Needless to say,
I never rode on one again. He had watched so many young people

die. I can only imagine the heartbreak that must have been for him and his fellow firefighters. How did they deal with it? One of my most poignant memories, though, was after an apartment fire where three children died. Dad found them. He wasn't Dad for months."

Harroun was two years old when Ron Brown had the close call that resulted in his friends saving him after a ceiling collapse. Her sister, Beth Mader, was six years old and discusses the impact that near-miss has had on her.

"This, for me, was an example of how every day in the life of a firefighter can go either way," says Mader. "Yes, Dad hurt his leg, which, by the way, predicted when it was going to rain for the rest of his life. He was off-duty for three months, but it also showed how close calls are every day. A falling fridge and stove from the above apartment in that fire missed Dad by too close, and [it] could have been his life then, instead of just hurting his leg."

Mader confirms that it was, in fact, a falling fridge and stove, not a washer/dryer, that crushed her father's leg, trapping him inside the fire. The family vehemently believes it was likely the materials the smoke-eater had breathed in while trapped that eventually led to his death at age seventy. Firefighters are often repeatedly exposed, over an entire career, to asbestos and other hazardous materials and chemicals.

A private man to the very end, Brown did not tell many people he was sick. His death took many firefighters by surprise, according to Harroun.

The Canadian Press

The loss of children is the focus of the last entry for this chapter. A January 2013 article in *The Chronicle Herald* cites the location of this emergency as Stittsville, Ontario. The article reads, in part:

"Police have identified the victims – two of them children – of what they describe as a 'horrific' double murder and suicide that has rocked a sleepy bedroom community on the outskirts of Ottawa. Ten-year-old Jon Alexander Corchis, six-year-old Katheryn Elizabeth Corchis and Alison Constance Easton, 40, were all residents of the two-storey home where their bodies were discovered Monday, police say. No specific details about the cause of death were released, although a police news release described the incident as a 'double murder and suicide,' and noted no charges were anticipated 'due to the circumstances.'"

"'It's horrific,' Ottawa police Insp. John Maxwell told a news conference. 'It's everybody's worst nightmare because it's so sad. It's a criminal act, but it's on the human-tragedy side of the balance.' ... Maxwell said police are speaking to friends, family and those in the neighbourhood to find out anything they can about what might have caused the tragedy. The family's past contact with police was limited to what he described as 'barking-dog calls.' None were related to domestic disturbances, he added. 'They certainly were not; absolutely not.' ...

"Maxwell also offered his condolences to the father, and expressed support to the firefighters and emergency crew members who were the first on the scene. 'I want to say that it was actually fire and EMS who arrived first, before the police ... the first officers on the scene, most of them have families; this is very, very difficult,' he said. 'Thank God there are men and women out there who go into the darkness like we do.'"

Chapter 9

Tracking Them Down

There was a lot of darkness for my family in 2006. I lost the one person who was my biggest supporter in life. After my father died on May 2, I had an opportunity several months later to appear on the now defunct, popular Halifax radio show *Hotline*, hosted by Rick Howe.

Howe is a great journalist with a legion of regular listeners on another station now, and is a friend of mine. His wife, Yvonne Colbert, is also a highly respected veteran Nova Scotia journalist. Colbert is one of my best friends and a long-time mentor and supporter. In fact, in 1987, when I first began in television, at CTV Atlantic, formerly ATV/ASN, it was Colbert, along with Bill Jessome, who were among a handful of broadcasters who helped nurture and guide me as a novice in the television news business. Jessome has since become an accomplished author and is well known in writing circles. He and I remain close friends. In his eighties now, Jessome still writes and has been a key force in convincing me to actually sit down and start writing this book.

In the summer of 2006, Colbert was filling in as the host of *Hotline* while Howe was on vacation. She invited me to come onto the show to discuss my father's passing, his rescue and Medal of Bravery. The aim or focus of the interview was to help me find the family of Nick McKenzie, a dream that started to take shape after my father's death. It was only weeks after his passing that I went on the radio show, and it was difficult for me to talk about him in public so soon after.

I asked *Hotline* listeners to call the show and provide me with any information they could about where Nick McKenzie or his family may be living. One person called in who knew someone, and so on. The day of the fire, there were hundreds of people who had gathered to watch. Many people saw my father enter the rear bedroom window and later emerge with the baby.

That first day on radio was exciting because I had made some progress; I knew I had gotten people thinking and talking. No one from the family dialed in; neither did Nick. I had no idea if they were still alive, still in Halifax, or whether they would even want to talk to me. At the close of my appearance, I asked listeners to keep the questions and stories going off-air. I provided the radio station my contact information should a lead occur.

One did.

I am not exactly sure what played out behind the scenes. But several weeks later, on the evening of September 5, 2006, I got a call at my home. I was alone in the kitchen. As these moments often happen, it came unexpectedly. The voice on the other end of the line was female. I did not recognize it.

The caller identified herself as Nick McKenzie's mother, Kim Gibson.

In shock, I sat down at my nearby dining room table. I could not believe it. Many emotions swept over me as Gibson and I spoke – excitement, happiness, relief and sadness. I was upset because I knew I would never be able to reunite my father with her or with Nick. That opportunity had passed. But here was another chance: a chance for me to get her side of the story, first-hand, as an eyewitness, and with the important perspective of the family.

During that first phone call, Gibson briefly recounted the events leading up to the fire and what happened during and afterwards. She kept referring to my father as "firefighter Landry." It made me smile and ache.

Safe to say, I did not do a great job of speaking with her that night. I took no notes and did not book a meeting, except I recorded on a small piece of notepaper the exact date and time of the call. I was overwhelmed with hearing her voice and finally getting to speak with her. I thanked Gibson and hung up. I burst into tears and laid my head down on the dining room table.

At that time, I was grieving and was not ready or able to tackle the lengthy research, interviews and writing that are involved in telling this story. That is why I did not pursue Gibson or McKenzie in 2006. I just could not do it.

I felt happy I had made a connection and knew she was still living in Halifax. She also told me Nick was away working in western Canada. At least I knew he was alive, happy and healthy. It was a rewarding phone call, in many ways.

Time passed. I tried my best to get back to a normal routine as a wife, mother and daughter. I pressed on with my work as a freelance journalist and instructing at Mount Saint Vincent University. I still do both jobs today.

In the summer of 2007, a little more than a year after my father died, I was invited back on the same talk show with Colbert. This time, we were joined by Bill Jessome, whom Colbert had also invited on air to talk about his stellar broadcast and writing career. It was fun in the radio booth with two of my dearest friends. After I set up the backstory of the rescue, I read "live" from a few of the historical documents I owned and invited callers to call in and speak with me.

They did, and this time the McKenzie family came out in greater force.

I discovered Nick McKenzie was living in the Halifax area with his partner at the time and his two children. The partner called in and spoke to me, as did his mother, Kim Gibson, and two of his aunts, including Charlene McKenzie Meade. They all tried to explain their gratitude and how Nick had grown into a loving and caring father. I didn't get to speak to Nick that day, or any time on the radio or phone. I made the assumption he preferred to keep his privacy. I tried to understand. I was a complete stranger and a journalist.

After the show, I spoke to his mother a third time. Gibson called me and said a family portrait was being taken and would be presented to me at a later date.

The show was a smashing success.

It was emotional for the listeners, presenters and staff. One person who was especially touched by what she heard was show producer Amber Leblanc. Leblanc sat in the control room side of the broadcast booth doing her job during the interview. In 2007, she wrote a blog that she posted online. Colbert sent me a copy of it, which I have kept. Here is what Leblanc wrote from the show day:

"From Amber's Blog: Hotline Radio Show, 920 CJCH, August 14, 2007

"Wow. What a show today. The first hour was so incredibly compelling. I was awed by Janice Landry's story about her dad rescuing the baby from the burning house last year [referring to show appearance number one] and to hear the conclusion to the story today was so gratifying, it brought tears to all of our eyes in the studio! The boy's mom, Kim, called the show and she was in tears too, remembering the day she was screaming on her front lawn as her baby was upstairs and no one could help him …

"It was devastating listening to this mother reveal how she felt when she knew there was a chance her son might die. It was equally as compelling when she told us how she felt when she saw her son and realized he was alive! As Yvonne mentioned, 'It's pretty obvious Nick was meant to live.' A little, tiny baby, with small lungs, survived a massive house fire and his mom told us he was not injured, he just had some soot in his nose!

"It was so great to hear that, because of Janice's plea last year, she found Kim, Nick and the rest of the family, and was able to make that connection. Nick's two aunts called too, to speak with Janice, and Yvonne spoke with Nick's wife after the news. Janice's dad sounded like a wonderful man, who even gave Kim an envelope with a little bit of money afterward and told her to go buy him [Nick] a lollipop. Nick is now married with a family because of what Janice's dad did. I can't tell you how incredible it was to hear a happy ending on the Hotline. It was also nice to hear the family refer to him as 'Fireman Landry.' It's kind of old school, like when we used to have heroes, and people had respect for people in uniform. Landry was, indeed, a hero."

Leblanc is bang on during her closing lines; the purpose of *The Sixty Second Story* is to pay homage and respect to people in uniform who do not always receive it. However, she was wrong about me finding or meeting Nick. I still had not connected with him this second time. I was now sure that meeting would never happen. I had figured, since I had spoken with his then partner, mother and aunts, that he had chosen to remain in the background. As disappointed as I was, I had to respect his decision and privacy.

I left it alone for five years, until the timing was right and I was ready.

In 2012, I began to plan what I would do to mark the thirty-fifth anniversary, in 2013, of my father's rescue. I knew, as a journalist, to look for a "hook" – an angle, number, person, fact or statistic that makes the story newsworthy and fresh. An anniversary is always a perfect time to let the public know about something newsworthy. But I also had to further the story.

I had to find and meet Nick.

I started to work on that and this book in January 2012, five months before the sixth anniversary of my father's death. I also wrote, simultaneously and strategically, a number of magazine articles on first responders for *Halifax Magazine*. It was the first series on a subject ever published in that magazine. The editor, Trevor J. Adams, is one of the people who first suggested and encouraged me to write this longer account.

To accomplish both projects, I knew I had to speak with Kim Gibson again. But this fourth time around, I wanted and needed a personal meeting, an in-depth interview and the chance to finally meet Nick face to face.

It was a lot to ask.

I knew the research would be arduous and taxing. Adding to the complexity of the project, five years had passed since our last conversations, so the old contact information I had for everyone was no longer useful.

I had to start over.

I went with my gut.

After interviewing Joe Young on May 14, 2012, I was buoyed by his vivid recollections and descriptions of what had happened. It was his interview that led me to a research breakthrough.

After our meeting, I ate lunch thinking about what Young had just told me and I sat there trying to figure out what to do. Instead of thinking, I acted; I went back to the investigative basics I had learned while studying for my honours journalism degree at the University of King's College, under the guidance of professors like Stephen Kimber and the late Ian Wiseman, who remain two of my favourite teachers of all time.

I started knocking on doors.

I got in my car and drove, for the first time ever, to the scene of the fire. It was twelve days after the sixth anniversary of my father's death. It is always a difficult time of year.

Colbert, who has encouraged me from the outset, also suggested I journal about my experiences. She was right, once again. Here is an excerpt of what I wrote about that day:

"After leaving lunch, I had a strong urge to drive to the location of the fire. I followed firefighter Joe Young's direction and drove down Bayers Road ... I drove into Federal Avenue and around the loop looking for the house. I passed two fire hydrants – it is the second one that Dad and Joe Young stopped at, in Rescue #2. I could not see the back of the home well. What Dad 'saw' that day is a mystery [Young is adamant Landry saw something from this vantage point that led him to immediately run to the back of the home]. As I approach 3390, I see a woman go into the front door. I decided I would call at the home and ask to see the inside. I was nervous, excited and went for it. It took a long time for the door to be answered."

I have cold-called on many doors as a former television reporter. Experience tells you this is risky business. You never know what or who awaits you. I have had a wide variety of reactions, including door-slamming, name-calling, personal threats, dogs barking, you name it. Most times, people are polite and simply decline your offer to talk. But there are always some who invite you in; persistence pays off in journalism – an invaluable lesson I had learned at King's from Ian Wiseman.

The journal entry continues. "The mail slot was pushed open and a child looked out at me. The door was opened by an elderly woman clothed in traditional Muslim attire. She did not speak English. I asked the little girl, the child beside the woman, if her mother or father was home. I was invited to come into the entryway.

"I am now standing inside. I do not know how much time I will have to size up the home and the scene of the fire. The staircase that neighbours and firefighters had initially tried to ascend in the thick smoke and heat is immediately to my left. There are about one dozen wooden steps climbing up to a second floor. I felt awkward and did not want to alarm or scare the people. I peeked around the corner into the living room. It was almost empty except for some beautiful prayer rugs on the floor.

"The mother eventually comes down the stairs. She is the lady I saw going into the home when I pulled up. She had been lying down because I could hear the bed and floor creak as she got up to see who was downstairs. She came down to greet me. I explained that my father had rescued a two-month-old baby from her home thirty-four years ago, and that I had always wanted to see the room where he almost died. She didn't hesitate to let me in. She did not speak a lot of English to me. We walked up the stairs. Her young daughter led the way. The hallway at the top still has several bedrooms branching off from it, like in the 1970s. The one in question is two doors down on the left."

The layout is exactly like my father sketched during the fire investigation. The bedroom where the rescue took place is very small, much smaller than I had pictured or anticipated. This reaffirmed the importance of my personal viewing of the location and not relying solely on the reports and interviews conducted.

The journal entry details the interior of the room: "It has a small bed with a duvet. The bedspread is covered with images of the children's character Diego, from the kids show *Dora the Explorer.* The room held only a bed and a dresser. It was sparsely decorated. There is one window."

The window is the one I have longed to see, my father's exact entry spot as he swung up and inside. A bed is now placed perpendicular against a wall with the window beside it. I stand there trying to imagine what it was like for my father in the black smoke.

I look out and see an overhanging porch roof, which now holds a satellite dish. The trellis is gone. There are still stairs leading into a rear entrance. There is a fresh load of laundry hanging on a clothesline to dry.

The little girl, whose name I do not ask for, is five years old. It is her bedroom, she says. She also loves *Dora the Explorer.* She opens up to me more than the mother. The grandmother remains downstairs but was always smiling and friendly. The mother tells me they have not lived there for very long and that she has two other children who are not home at the time. The little girl carries a doll with her. When I ask who the doll is, she replies, "Baby." It is emotional to be speaking to a child in the very room my father rescued another baby from thirty-four years ago.

The little girl goes into another bedroom, near the top of the stairs, to show me another favourite of her toys: My Little Pony. She is adorable and a perfect tour guide. I do not stay long. I do not want to leave, but I also do not want to overstay my welcome.

As I reluctantly descend the stairwell, I count the steps. There are, indeed, twelve.

I thank the mother and ask if I can drop presents back to the children as a thank-you gesture. I tell her my name is Janice. She agrees to let me return, but I can tell she is hesitant. I tell her I will be back. As I walk to my car, I turn around. She is watching me, puzzled and inquisitive.

I smile, wave and drive away.

I do not get far before I stop. I park the car around the corner. I want to see the home from the rear without alarming her.

I see a woman in a nearby parking lot who is unpacking her groceries. She has two large bottles of juice in her hands. I tell her my father was a firefighter who rescued a baby from a house and I point to it. I ask her if anyone still lives in the area from the 1970s who would remember the fire. Without hesitating, she says, "You have to talk to Pops." She stops what she is doing and personally escorts me to a house nearby. She rings the doorbell and knocks on the door. Once I explain why I am there to the woman who answers, the other lady leaves.

The woman who opens the door is Judy Mamye. I tell her who I am and she immediately ushers me in. A man is resting on the sofa. He is her husband, David Mamye, but everyone calls him Pops. I explain that I have just come from the scene of the 1978 fire. Judy vividly remembers the day. She says her son, also named David, and a friend were sitting on their back deck when the fire started. The two young men, and some other guys from the neighbourhood, rushed to get a ladder. Judy says the smoke was thick and pouring from the home.

To my amazement, she tells me that Nick McKenzie's younger brother, Corey, has just been in their home the night before. She says he drops by fairly regularly and has lived in Halifax his whole life. Judy takes my business card and offers to try to help me connect with him or Nick. She says Nick is shy. I had already concluded that, but I am now excited because a family friend is helping me track down Nick and/or his brother.

Pops Mamye says he knew my uncle, Lawrence Landry, my father's brother who was a Dockyard firefighter. Pops worked as a carpenter at the Halifax Shipyard and made furniture for oil rigs. Their tidy home has several beautiful pieces Pops has made, a grandfather clock in the living room and a hutch in the dining room. Pops explains that the interiors of the units on Federal Avenue have changed somewhat since the 1970s. I am glad I have now been inside two.

Pops' own father was also in a fire department on Prince Edward Island. The number of firefighters or relatives of firefighters I have encountered during my research keeps right on climbing. Is it coincidental that it is mainly firefighters or relatives of firefighters who are trying to help me – the daughter of a firefighter?

I do not think so.

As former chief Don Swan prophetically said, "… this is all family."

The Mamyes confirm most of the neighbours who were involved in 1978 have long since moved away. They are the only ones left of the original occupants still living close by. My strong urge to simply drive over to the location, on that particular day, at that particular time, has paid off; the lady who was unpacking her groceries led me to the Mamyes and one step closer to Nick McKenzie.

I conclude my journal entry by writing, "I leave with the hope they will be my final connection in my efforts to track down Nick. I am blown away by the generosity of all these strangers."

Nine days later, everything started falling perfectly into place.

Chapter 10

Meeting Kim Gibson

Prior to Joe Young's May 2012 interview and my visit to the scene of the fire, I also started cold-calling a number of MacKenzies in the phone book, trying to find the family involved in the 1978 fire. Notice the spelling I had been using: MacKenzie.

Part of the problem complicating my work and research was the fact, as you have now read for yourself, some fire reports and media stories from the 1970s and 1980s had documented Nick's first and/or surname incorrectly. It is actually spelled McKenzie, not MacKenzie.

That one error has hampered my search.

Since the old-fashioned phone book approach was not working, I turned to its more modern counterpart, social media. I have Twitter and Facebook accounts and use both daily. I decide to go online and start typing in various spellings of MacKenzie and McKenzie in the search box on my Facebook account to see who pops up.

No one with both names matching does, at first.

But eventually a man named Nick McKenzie appears. Looking back at me from my home computer screen is the somewhat familiar face of a grown man who bears a resemblance to an old photo I own of the baby with my father. I try not to get too excited.

On April 23, 2012, at 3:06 p.m., I send the following Face-book message to this stranger: "Nick: not sure if I have the right person. My apologies if I don't. Looking for a man with your same name who was rescued from a fire as a baby by a Halifax fireman. I am the fireman's daughter. I also sent your mom a message. Trying to track you guys down as I am writing a piece about dad for next year. Thanks, in advance, for your help."

I wait.

Days pass.

I start to get disappointed.

But on May 1, 2012, at 10:35 p.m., I get this response from the man named Nick McKenzie: "Hi. You did find the right Nick. I'm very grateful for what your father has done for me. If there is anything I can do for you and your family, please let me know."

One day before the sixth anniversary of my father's passing, I have finally found and spoken, albeit briefly, with Nick McKenzie. Thirty-four years after his rescue, McKenzie is now Facebook "friends" (the term used when someone accepts your invitation to connect on the social media site) with the daughter of the man who had saved his life.

Nick's willingness to befriend me on Facebook allows me to view photos of his two children, a boy and a girl, read posts about his career, and view other family names and information. It was a generous gesture on his part to let me into his world. Because we are now connected online, he can see my pictures and posts too.

Now that I can find other family names on Nick's Facebook wall, I am also able to connect with his mother, Kim Gibson, and aunt, Charlene McKenzie Meade, through social media. Meade was one of the two aunts who had originally called into the *Hotline* radio show in 2007. Both women also readily accepted me as "friends."

According to his Facebook page, Nick was living and work-ing in Fort McMurray, Alberta, at the time of our connection. Now I know I have to be patient in order to ever meet him in person. But there is hope. During that time, Nick would fly back and forth from Alberta to Nova Scotia for family visits.

It is my hope to catch up with him on one of those occasions.

Behind the scenes, I later learn, the Mamyes had kept their word and were still trying to help me. Eventually, all communication roads led to me speaking, once again, with Nick's mother, Kim, on the phone. We agree to meet in person, for the first time.

I ask for a formal interview and am finally prepared to find out exactly what had happened, in detail, from her perspective. I meet face to face with Gibson at her Halifax home on May 23, 2012. She is able to fill in many blanks the retired firefighters and official reports cannot because they were not present before and as the fire started.

In fact, her memory from that day is razor sharp; she speaks with great conviction, emotion and clarity. Driving to her home I am excited and nervous. She greets me warmly and invites me to sit in her living room.

She is petite with light brown hair, which falls around her shoulders, has a slight build and is friendly. She apologizes for not having pictures of her two sons hanging on the walls. She says she is waiting for one of them to come and arrange them for her. There is a small pot of flowers and a Mother's Day card from her younger son, Corey, sitting in plain view, on a side table.

I stay for two hours.

It was an incredible conversation and interview.

"You ask me something that happened last week and I might not remember it. But you ask me about that day, and the hairs will stand up on my arm. I can remember that day as if it was yesterday. I can see your father's face. I know your father's face," is how Gibson opens our conversation.

Gibson's comment is oddly reminiscent of how the firefighters are able to recollect scenes from decades ago. I do not have to ask a lot of questions. Once she starts talking about October 1978, the words just pour from her. "I lived with Mom and Daddy because Nicki's father [she always calls him Nicki] passed on when I was four months carrying my son."

Nick was two months old at the time. This means Gibson was still in the midst of grief coping with the recent loss of her partner and Nick's father, Nicholas Downey, when the fire occurred.

Gibson confirms the home's layout, which is exactly like reports I have obtained. "When you walked in the front door, the upstairs was to the left. You had four bedrooms: the first room on the left was the bathroom, then there was my bedroom [where the rescue occurred]. On the end was my little brother's bedroom [where the fire originated]. I had two sisters who slept in another bedroom, and then my mother and father's room [closest to the top of the stairs, facing the front of the house].

"That day my older sister, Charlene, her and her husband, Danny, who was in the service [military], they were just getting back from Greenwood. Danny was working. Charlene was there with her daughter, Shannon, who is two months older than Nicki. She had Shannon downstairs in a playpen. So what happened that day is: Mom's washer broke down. I asked my other sister, Nanette, 'Can you take me over to the laundromat to wash Nicki's clothes?' My brother, Jamie, was eight years old at the time. He came with me and Nanette. Mom looked after Nicki."

"When I came back in, Mom said, 'Don't worry about Nicki,' because I was always overprotective. I always had the bedroom door open so I could hear him. 'Oh, he's sound asleep. The door's shut. Leave him alone,' Mom said."

Gibson says her sister, Charlene, had been cooking hamburgers for the family in the kitchen. She invited Kim to sit down so she could eat one. Before doing so, Gibson says she placed two garbage bags filled with Nick's clean clothes in the home's front entryway. Because she did this, Nick was the only family member afterwards who had any clothes spared from the fire.

Gibson reveals the exact moment she, and then the family, knew the house was on fire. "So I was sitting there with everyone, including my cousin [also named Kim], who was at the house. We were sitting in the front room. Everything was quiet while people were eating. We never had a fireplace, but I heard this noise, cracking, like how a fireplace crackles. I said, 'Charlene, did you leave the frying pan on?' She said, 'No.' I said, 'Well, I hear something crackling.'

"So I got up and came out of the living room into the kitchen and I saw nothing. I went and turned on the hall light. All I could see was rolls, just rolls of black smoke coming down the stairs. I just screamed and said, 'Oh my God!' Daddy jumped up and he tried to get up the stairs, three times. I tried and slipped down the stairs and I broke my thumb and wrist. I remember something so clear. My father looked right past me to my mother and ..."

At this point Gibson breaks down.

Struggling to compose herself, she continues, "I never saw my father in my life cry, but he said 'Shirley, I can't get him.' At this point, flames were shooting up from the stairs. The next thing I remember is being out on the lawn and screaming. People were just coming from everywhere, trying to put up a ladder, to get upstairs. I remember a little man who came from across the street and tried to get up the stairs with my father. The man came down full of black smoke. He couldn't get up. All I remember is sitting on the lawn, rocking back and forth screaming."

It took three minutes for the fire department to arrive.

"I didn't even hear the fire trucks come up. All I remember is being out on the lawn and somebody grabbing me and saying, 'Come around back! There's a fireman. He's getting your son!' And when I went around back, I saw your father leaning out my bedroom window, with my baby in his arms, in a blue blanket. He was kind of gasping for breath when he was leaning. Another fireman was just then putting the ladder up and your father passed Nicki to the other fireman.

"When they came down, I swear there were hundreds of people all gathered around. I was still screaming and crying when he passed Nicki to the other fireman on the ladder. Nicki's eyes opened and everybody went, 'Ah!' They clapped. That's when we got into the ambulance and we went down to the hospital. They said Nicki had no smoke inhalation and that fireman Landry had given him mouth-to-mouth. I had to take Q-tips and take soot out of his nose. He had a little difficulty taking his milk. The next day, he was fine."

That next day, my father returned to the fire scene. It was there that he spoke to the family for the first time, according to Gibson: "Your father told my Mom, 'It's a good thing that your husband or nobody got up through those stairs because they would never have survived coming back down the stairwell.'"

Gibson explains what the aftermath looked like. "All the other rooms were burned right out, right completely. My bedroom ceiling had bubbles, black bubbles, like it would have combusted anytime. Your father also said to Mom, 'You know, Mrs. McKenzie, you literally saved his life by keeping that door shut.' I always would have had that door open. Mom had Nicki on his stomach with a blue blanket over him."

The blanket, and the door being shut, had helped to shield the baby from the thick smoke.

"All the smoke was coming through the cracks of the door. I'll tell you it was a matter of seconds! I know that for a fact," she says emphatically of how important the timing was in this rescue. As the facts stand, and by every account, it was an extremely close call.

"I'll tell you, Janice, the way I look at it, if God wanted to take my son, he would have taken him. He let your father save him for a reason. So all I can say is, 'Thank God.' He was an angel that God just sent there to save my baby. He is almost thirty-five years old and when I talk about it ..." She breaks down again and pauses, before saying, "... it's just like it was yesterday."

Even though it was a positive outcome and no one was seriously injured or died, the what-ifs have plagued Gibson. It is not just first responders who suffer emotionally and physically after traumatic experiences. The victims do as well. As FFAP coordinator Paul MacKenzie explains, the lasting effects vary from person to person.

"I ended up with a medical problem. I ended up with OCD [obsessive compulsive disorder]," says Gibson. "Every night I put my kids to bed, I would check on the closet over and over again. I couldn't sleep. I'd lie down. Three minutes later, I'd get back up. I would check the kids. Then, I would put my hand on the stove. I would constantly do that, over and over again. I moved to Toronto and met my second husband. He noticed and suggested I see a doctor. He said it was due to the stressor of the fire."

The OCD is much better now but is not completely gone. "I don't have to be to work until seven-thirty in the morning. I go there at six-thirty. The reason being? I have to clean my computer. I have to have everything set, just right, before I start work."

Gibson still checks the stove before she leaves for her job at a Halifax call centre. As the daughter of a firefighter, I admit I do the same.

Gibson has had her fair share of heartache. She has had to deal with the loss of Nick's father, Nicholas: "I was nineteen when Nicki was born. I was very young. Nicki's father was my first boyfriend. We went together for one and a half years. We had planned to get married. When he died, it was the worst thing because I had never had anyone die in my life. I was with him that night and he was on his way home. He fell through the ice and was missing for nine days. It was a horrible search. It was horrible. Mom said I cried for about six months. I saw a priest who tried to explain death. I came through it. Nicki looks like his father."

After those six months, Nick is born. Two months later, the fire happens. On top of that, her second husband, Danny Gibson, whom she lived with in Toronto, died of bone marrow cancer. "Mom says to me, 'You are the strongest member of this whole family because of everything you went through.' That fire almost took me over the edge. I had my son sleep with me until he was eight years old," she says, laughing a little. "It wasn't until we moved to Toronto and I was with my husband, Danny, that Nicki slept in his own bed. Nicki was my life, and today, he and my other son are my life. I am always worried –" She breaks down for the third time, "– about Nicki, because the job he does is so damn dangerous."

Gibson says her late father, Jimmy, who died in 2002, was a firefighter in the Royal Canadian Air Force (RCAF). Her sister Charlene's husband, Danny Meade, was also an RCAF firefighter, who rose through the ranks to become fire chief at CFB Toronto. That means the man who initially had tried to make it up the stairs to save his grandson was a trained firefighter who would have realized the precariousness of the situation. Meade was not present when the fire broke out. He was at work.

"My mom was screaming and crying for about two days later after the whole shock of everything. I ended up staying in Hatchet Lake with my mom's sister and husband. We lost everything. They had a cradle that we brought out there. I stayed with my cousin until they repaired the place. Your father came back the next day to go through the whole house and everything. That was when Mom was crying and your father was consoling her. There are no words that you can say to a person who saved your son. How can you tell a person how much you're grateful? There are no words to possibly express that," Gibson concludes.

As difficult as it is for her to sum up its magnitude, the trauma stemming from 1978 clearly still exists. Despite that, the McKenzie family has graciously and openly discussed the terrifying day for the benefit of my father and my family. It is also difficult for me to express my gratitude for their collective candour.

Gibson and I end our lengthy interview. She agrees to try to set up a meeting with Nick when he returns from Fort McMurray. The spring and summer of 2012 pass without any further McKenzie family interviews or meetings. Instead, I focus on speaking with the firefighters featured in this book.

Months pass. I start my first responder magazine series and I am thrilled with its reception. The thirty-fifth anniversary of the rescue is looming. I know I need to meet Nick to be able to pull this whole thing off.

It is almost the end of 2012, December 7, to be exact, when I finally speak with Gibson's sister, Charlene McKenzie Meade, via telephone. We had been trying to meet in person for weeks but the timing was not cooperating. As the year was drawing to a close, I want her viewpoint and quotes nailed down so we decide to speak over the phone. McKenzie Meade was the family member frying hamburgers just before the fire broke out.

She was twenty-two in 1978.

McKenzie Meade says she retired in 2012, after working twenty-five years with the RCMP. She was the force's personnel security manager for the Atlantic Region. McKenzie Meade says her husband, Danny, had been working as an RCAF firefighter aboard HMCS *Ottawa* at the time of the fire. She explains the ship had helicopters stationed aboard, which required the first responders to also be at sea.

There was another fortuitous twist of fate, besides Nick's grandmother, Shirley, closing the door while he was sleeping, which was never typically done. McKenzie Meade's daughter, Shannon, who was just two months older than Nick, was supposed to be sleeping in the cradle in the very room where the fire broke out. Her baby daughter was not settling down well so she kept her downstairs in a playpen while she cooked the hamburgers.

McKenzie Meade remembers the fire making a whooshing sound and that the "heat was so intense." She ran out the door with Shannon and called the fire department from a neighbour's home because she says the power in all the attached row-housing units went out.

"The firefighters seemed to take forever to get there. The first ones ran into the building without breathing apparatus on but were forced back out immediately to get their air packs."

McKenzie Meade says she is the person who told my father there was a baby trapped upstairs. She also showed Landry exactly what room Nick was inside. "I showed him and then saw him starting to make his entrance and then I returned to the front to look after Kim."

She vividly remembers the huge crowd outside clapping and hollering when Nick was brought out. "I went to the hospital with Kim and Nick to have Shannon checked over and to help Kim. Shannon was unharmed."

Yet it had come dangerously close, once again, to it being a tragic outcome, not only for Nick but also for his cousin. "The cradle melted in the room. That is where Shannon would have been," she says.

Would have been – the would-haves, could-haves, and what-ifs in this story keep recurring.

There was, in fact, a cradle in the drawing my father made of the room where the fire originated. It has never before been revealed, before McKenzie Meade's interview with me, that another infant, a baby girl, could easily have died in it.

Gibson's sister is, in fact, the person who finally helped me arrange the long-awaited meeting with her nephew, Nick. I actually found out on Facebook about her behind-the-scenes effort to make this happen. She created a new Facebook event called Meeting Nick McKenzie and Janice Landry. She invited Kim Gibson, Nick McKenzie, Shirley McKenzie and other family members. The notification and invitation was sent to me two days after our interview, on December 9, 2012.

She had unknowingly decided to schedule the meeting of a lifetime for Friday, December 14, 2012, two days before what would have been my father's eightieth birthday.

Chapter 11

Meeting Nick McKenzie

After the invitation goes out on Facebook, I have to wait six days for the visit with Nick McKenzie. I find out I will also be interviewing his grandmother, Shirley McKenzie, at whose home the fire occurred.

The event will be at Gibson's Halifax apartment. I sleep poorly the night before. I have butterflies in my stomach the whole morning. I drive over feeling nauseous and certain that my long-awaited meeting will not actually occur; I have waited so long and when you are a journalist, until a meeting is done and an interview is complete, the process is far from over. People constantly change their minds and back out of interviews, especially when it is an emotional subject they are discussing.

At Gibson's home, I am the first to arrive.

The others run late.

I try to be patient, which is not a trait that comes easily for me.

Shirley McKenzie arrives before her grandson. Accompanied by her daughter Leanne, who was not present at the time of the fire, she seems happy to meet me. The feeling is mutual. Leanne takes a lot of pictures over the course of the afternoon and avidly reads all the fire documents I bring along that are part of my father's collection.

Her mother, Shirley, is a beautiful, mature lady. She is wearing a black sweater with jewels embroidered on it, a necklace, earrings and brightly coloured lipstick. I invite her to sit down next to me on the sofa. She instead chooses a chair next to it, which she says is always the chair she sits in when she visits her daughter Kim.

I inch over towards Shirley so I can sit as close to her as possible, without making her uncomfortable.

"I'm too emotional to even talk right now," she says, after I show her a picture of Nick and my father taken the year after the fire. Nick is sitting on my father's knee and Dad is wearing his firefighter's dress uniform. Nick was about fourteen months old at the time.

Shirley says she was in her forties in 1978, like I am now. She has great difficulty discussing how her late husband, Jimmy, had repeatedly tried to save their grandson. It is upsetting for Shirley to revisit that moment. However, she does confirm Jimmy did, in fact, make it all the way up the twelve stairs and crawled towards Kim's bedroom but was forced to turn around because of the perilous conditions. Jimmy's face and body were black from all the rescue attempts. He also suffered smoke inhalation.

She became hysterical when Jimmy said directly to her, "'Shirley, I have to tell you something. We can't get the baby. There's no way.' When he said that, I screamed. I was out of it. I ran out screaming," she says, breathless and teary, remembering those initial, horrifying moments when she and Jimmy thought Nick was going to die in their home.

Shirley remembers that in 1978, Trina (Forbes) Crosby, the mother of Canadian hockey superstar Sydney Crosby, lived across the street from the McKenzies with her family, including several brothers. The Forbes family are one of the groups of neighbours who had tried to save Nick: "When I went screaming, 'They can't get the baby! They can't get the baby!' the Forbes – Sydney Crosby's mother was a teenager then, her brothers got a ladder and they came to the front," Shirley explains.

Shirley clarifies the ladder was mistakenly placed by the brothers at the front of the home and up to a window facing Federal Avenue. "They were up trying to get in my bedroom window. That was my window. I said, 'No. No! It's around the back!'"

By that time, the fire department arrives.

Shirley remembers seeing my father run around the back of her home, after her daughter, Charlene McKenzie Meade, had shown him where to go. Shirley is hysterical. Neighbours try to calm her down. "I thought the baby was dead. The next-door neighbours had to take me in because I was too out of it."

This is the same moment when Kim is also on the front lawn screaming for help.

A family named McCarthy, who had lived at the far end of the attached housing units, took Shirley inside their home to try to help calm her down. "When they came in and told me your father had got Nicki and he had been unconscious, and your father had given him mouth-to-mouth, everybody was clapping and going crazy. I wouldn't believe it. I said, 'No. The baby's gone! The baby's gone!'"

She says to me, "With his wisdom and his determination, he found a way. He would not stop. And when I'm looking at his daughter, oh my God, it's such a miracle! I'm going to be saying 'thank you' to your father for the rest of my life."

Almost as soon as Shirley finishes recounting her viewpoint and perspective, the interview I have waited to do for more than thirty years is about to happen.

The door opens and I hear a deep but quiet male voice.

I cannot see Nick at first because there is a wall dividing Kim's kitchen from her living room, where Shirley and I are seated. I have imagined what he looks like for a long time. Nick's Facebook pictures have helped me fill in the blanks.

Thirty-four-year-old Nick McKenzie walks into his mother's living room toward me. I have visualized this moment many, many times.

In reality this is what I do: I get up and hug him. Even though he is a complete stranger I feel like I already know him.

Nick stands about five feet nine inches tall, has a medium build. He is handsome with a shaven head. He has a big smile. He speaks quietly and in one-word answers at first. He must be embarrassed by all the attention. I have already been warned that he is quite shy. Before we start the interview, Kim and her sister take a lot of pictures of Nick and me, seated side by side on the sofa.

It is a meeting and an interview thirty-four years in the making. Over that time, I have been working as a journalist in Halifax for twenty-six of those thirty-four years; this is, by far, the most important interview I have ever done.

I am grinning from ear to ear the whole time. Listening back to my notes, I notice I have talked too fast, like I did when I first started in broadcasting when I was nervous or under pressure. I also have talked too much, like I do most of the time and especially when I get excited. I was clearly nervous. I did, however, manage to conduct a proper interview.

This is good because I do not plan to re-interview the family and ask them to revisit this painful time again.

McKenzie says he grew up hearing the stories about the firefighter who saved his life. I formally present Nick with one of the pictures I own of my father from the time of the rescue. It is a black and white portrait, shot outside, of Landry wearing h=is dress uniform and standing in front of Rescue #2, the white fire truck he arrived in that day in 1978.

On the back of the photo I have written: "To: Nick, From: Janice Landry, December 14, 2012: Thirty-four years after Basil (Baz) Landry rescued you from a fire at 3390 Federal Avenue. It has been a blessing meeting you." The photo has July 13, 1979 on its back.

I tell Nick I have had a hard time parting with an original picture because I own so very few of Dad during his era on the fire department. Nick responds, "Well, thank you. It's in good hands."

I imagine it now, hanging on a wall of McKenzie's home, Dad still watching protectively over him. The other firefighters involved in Nick's rescue have all wondered how he is doing. Now they know he is doing really well. In fact, Nick has gone on to have a

Nick McKenzie with his grandmother Shirley McKenzie, December 2012.
(Photo by Janice Landry.)

Kim Gibson with her son Nick McKenzie, December 2012. (Photo by Janice Landry.)

full and happy life, without any medical issues stemming from the fire. He was no long-term effects whatsoever.

He has two children: a son Mason, who, as of December 2012, is ten years old and, a daughter Alexa, who is eight. Nick pulls out his wallet and proudly shows me photos of his children.

Ironically, Nick's current vocation involves danger, an interesting career choice given the drama that had unfolded in his early childhood. McKenzie says he works with Remote Access Technology (RAT) based out of Dartmouth, Nova Scotia. He started with RAT in 2003, which has offices through the United States and Europe.

Nick works with his uncle, Danny Meade, Charlene's husband and former RCAF fire chief, who has been with RAT for fifteen years. Meade is the service line manager for Canada. Nick explains the type of work involved. "We do rope access. We repel to get into position. For example, we have worked on the smokestacks in the [Halifax] harbour."

Nick is describing the three tall red and white stacks which sit side by side on the Dartmouth side of the harbour. They are prominent fixtures on the Halifax metro skyline. He has worked on all three, examining the structures for safety purposes. "Any maintenance that needs to be done, we climb to the top and repel down the ropes to get into position and work on them, on the outside and the inside. I've been inside them. There's another stack inside. As you go to the top, it tapers in and gets tighter."

Nick says he is not claustrophobic but heights can still scare him. He has learned to respect them: "I do fear heights. When I get up to about five hundred feet, my legs start to shake a little. When I start working, it goes away."

Nick explains that the three smokestacks are, in fact, about that tall; the one closest to Bedford Basin is 505 feet high, the centre one stands at 515 feet, and the one closest to the mouth of the harbour is 525 feet in height. "We mainly do inspections on the outside, to make sure the concrete is still good," he says.

Then he reveals something of a bombshell. His work with RAT has led him to being involved in another close call, the second one in his lifetime. "I had one of the main [safety] lines on my ropes burn through. It was one and a half years ago, in Fort McMurray."

Nick actually works about forty-five minutes north of there, in an area called Wood Buffalo. He commutes back and forth, from east to west and vice versa, like thousands of people from the Maritimes, to earn his living. Nick was one hundred feet up in the air working on a steam pipe when his safety line separated. "The mainline broke because it was sitting on a hot pipe. I fell onto my back up line. I fell about two feet. We always use two [lines]."

I can only imagine what it felt like when the fall started at one hundred feet up and when he stopped after only two feet and realized his life had been saved again, this time by a rope.

I ask him what his reaction is to almost falling to his death. "I went into the smoke pit and had about a half a pack of cigarettes and went back to work. I had been complacent."

He says it matter-of-factly but with emphasis.

I know he has learned the hard, hard way about safety, way back when and much more recently. Now a ten-year veteran of rope access technology, Nick has worked on ships and in the oil and gas industry across Canada. A respected member of his team, he now teaches others about confined space entry and rescue. As of our late 2012 meeting, he has been instructing in his field for about five years. Landry would be proud of the hard-working father of two.

His work is his passion and so are his two beautiful children. Nick says he has even bought his son, Mason, a climbing harness for his birthday. The two McKenzie boys go wall climbing together. Nick tells me, as our time together wraps up, that his children are the joy of his life.

Meeting Nick McKenzie has certainly been one of mine.

Nick McKenzie and Janice Landry, December 2012. (Photo by Leanne McKenzie and Kim Gibson.)

Chapter 12

3390 Revisited

It has actually been two families from 3390 Federal Avenue who have made a big impact on my life: first the McKenzies and later an unknown family who welcome me in during the spring of 2012 to witness, first-hand, where my father's extraordinary rescue had occurred.

I had promised the woman who lived there that I would return with some small gifts for her children to thank her and her family for their kindness and generosity. She had given me a once-in-a-lifetime opportunity of climbing the twelve steps to the bedroom where Landry saved McKenzie.

Thanks to my dear friend Yvonne Colbert, I had been journaling throughout my research, which means I have another special entry about my final visit to 3390 Federal Avenue. It seems fitting to close with it and this next scene because it is there that *The Sixty Second Story* began and it is here that it will end.

My final journal entry reads:

"There is a black and orange cat sitting on a step in the front when I arrive. Some neighbours sitting nearby outside eye me suspiciously. They know I don't live around here. I ignore them. As a long-time, former reporter I am used to those glaring looks.

I knock on the door.

Her son answers.

I ask if his mother is home.

He asks me, 'Mother or grandmother?' I say, 'Mother.'

He calls out, 'Mommy, there is someone here to see you!'

She comes to the door. I say, 'I told you I would be back. I wanted to thank you for allowing me in to see the room.'

I have brought with me two bags of stuffed toys for her three children and a twenty-dollar grocery store gift card, along with a note card which reads, "Thank you for inviting me into your home, without hesitation, as a complete stranger, where my late father, a firefighter, rescued a two-month-old baby in 1978. To see the room is a true blessing. Thank you. Janice Landry"

I give the two bags of toys to the two children: one to the boy, who tells me he is nine, and the other to the little girl, who is five and sleeps in the room where the fire happened. They quickly haul out the stuffed animals. The little girl clings onto a tiny stuffed fish that has big pink lips. She kisses the toy. She is thrilled. I take a brown teddy bear wearing red flannel pants out of the bag and give it to the boy and tell him the bear is his and that they will have to share the rest.

The mother keeps saying to me, 'Why did you do that?!'

She also keeps repeating, using broken English, that when she told her girlfriend about me previously coming over, her friend had warned her to be careful because I was a complete stranger.

The mother concludes, 'You can't trust everybody.'

I tell her I agree.

I ask the woman to open the card in front of me.

I read her the inscription to ensure, this time, she fully understands why I came in the first place.

She is surprised when I explained my dad was a firefighter. That was apparently lost in translation during my first visit, which is why my reading the card aloud was imperative this second time round.

I also explain that dinner for the family was on me, thus the grocery store gift card. Again, she says, 'Why did you do that?!'

I don't reply.

There is no need to because she finally understands.

I thank her and reassure her I won't be back again.

When I get to my car I turn to look back at her.

She is standing in the doorway watching me, like she did the first time. But on that initial visit, her expression was a mix of apprehension, confusion and uncertainty. This time, the look on her face is dramatically different; she wears the most beautiful smile I have ever seen.

In fact, this kind stranger keeps right on smiling at me while I buckle myself in and drive away."

A different family lives at the house now, but like the McKenzies, they have welcomed me and warmed me with their generoisty. My heart is full. Thirty-four years later, there is, once again, something to be happy and grateful for at 3390 Federal Avenue.

Postscript:

Line-of-Duty Deaths, 1878-1980

Although the first chapter of *The Sixty Second Story* is dedicated to the nine firefighters who died as a result of the Halifax Explosion, I have chosen to include, as a symbol of respect, the names of all of the firefighters whose names appear on the Halifax Fallen Firefighter Monument. It should be noted there are other similar monuments across Nova Scotia and Canada. The process of reviewing firefighter deaths and determining whose name is to appear is sensitive, lengthy and complicated; criteria vary.

The official list of line-of-duty deaths (LODD) is carefully researched, compiled and maintained by current and former members of Halifax Regional Fire & Emergency, including former administrative captain and historian Don Snider; treasurer of the International Association of Firefighters, Local 268, Chris Camp, who is currently a career firefighter; and veteran Dave Singer, the driving force behind the Halifax Fallen Firefighter Monument. They are assisted by others who voluntarily serve on the committee.

Camp further explained in an e-mail, "We are the Halifax Fallen Firefighter Monument Committee. The monument went up before amalgamation and you are correct that we are now HRM. All LODD's on the monument are from the former City of Halifax. In the spirit of the original monument dedication [the 1917 explosion deaths] we have a strict requirement that the death has to occur at the scene or clearly as a direct result of injury incurred at the emergency scene. We have no others that meet these criteria at this time. There may be others recognized by other organizations with their own criteria but ours is strict for the reason stated above. We are hoping to erect a second, smaller monument recognizing occupational disease/illness deaths as there would not be enough room on the current monument."

Recognizing line-of-duty-deaths is highly important to every fire service. Snider wrote, via e-mail, to say he has also assisted a second group, post amalgamation, in erecting its own special monument. "The volunteers in Porters Lake/Lake Echo erected a monument on the station grounds, and from our former Halifax fund, which I had control of, I donated five hundred dollars towards it from the balance of funds of [the] former Halifax FF Monument Fund. Since provincial legislation passed several years ago, certain cancers causing deaths are also deemed LODD's. We talked about erecting side units to the monument in order to place the names of those members ... The Canadian Fallen Firefighters Memorial in Ottawa has a very broad description for LODD's and there are names of Halifax members who are not on ours but on theirs."

All of the late first responders, including my father, and the other fallen firefighters described in *The Sixty Second Story,* have devoted themselves to saving lives. Their legacies and stories are also important and have created lasting impressions; that is, in fact, the core message and meaning behind my work and this book.

With no disrespect meant for anyone, and with the realization this is highly sensitive subject matter, the following list is the official one approved by members of the Halifax committee solely for the monument at Station #4 on Lady Hammond Road, which began as homage to the nine killed as a result of the Halifax Explosion.

The list reveals, in total, twenty-four Halifax firefighters have been killed serving in the line of duty, the first recorded death occurring in 1878. The last one happened thirty-three years ago, as of 2013, in 1980. There were seven deaths in six separate incidents, before the nine men were killed in the Halifax Explosion. There have been eight deaths, in eight fires, since 1917.

There has not been a Halifax firefighter killed in the line of duty since the former Halifax Fire Department officially became the current Halifax Regional Fire & Emergency. This is the verbatim list and individual descriptions which I sent via e-mail to Chris Camp and others for review and fact-checking:

Line-of-Duty Deaths

Lieutenant Edward Fredricks, age 26: died April 14, 1878, while fighting a fire on Cornwallis Street. There was an explosion within the building causing it to collapse. It took some time to find him and another firefighter who miraculously survived.

Hoseman Rufus Keating, age 32: died January 19, 1894, following a fall from a ladder during a fire at W.M. Stairs, Son & Morrow on Lower Water Street. He landed on a steel fence post, which pierced his throat area and caused him to choke.

Lieutenant William Lewin, age 29: died March 4, 1898, when a chimney collapsed during a fire on Tower Road, causing him to fall from a ladder.

Hoseman Michael Sullivan, age 48: died March 6, 1903, from exposure and smoke inhalation during a major fire at Moirs factory, Duke Street.

Hoseman Richard Supple, age 29: died March 21, 1903, from exposure as a result of a major fire at Moirs factory, Duke Street.

Hoseman James Tynan, age unknown: died June 9, 1909, during a major fire at the Halifax Furnishing Company, Argyle Street. The building collapsed, injuring seventeen, trapping Tynan, causing his death.

Hoseman/Driver William Knapman, age 25: died April 15, 1915, following a fire on Argyle Street. He was dragged up Buckingham Street when one of the horses broke loose. He struck his head, leaving him unconscious.

Fire Chief Edward Condon, age 60: died December 6, 1917, during the Halifax Explosion.

Deputy Chief William Brunt, age 41: died December 6, 1917, during the Halifax Explosion.

Captain Michael Maltus, age unknown: died December 6, 1917, during the Halifax Explosion.

Captain William Broderick, age 32: died December 6, 1917, during the Halifax Explosion.

Hoseman John Spruin [his age is not listed but he was 65]: died December 6, 1917, during the Halifax Explosion.

Hoseman Walter Hennessey, age 25: died December 6, 1917, during the Halifax Explosion.

Hoseman Frank Killeen, age 21: died December 6, 1917, during the Halifax Explosion.

Hoseman John Duggan, age 34: died December 6, 1917, during the Halifax Explosion.

Hoseman Frank Leahy, age 35: died December 31, 1917, following injuries from the Halifax Explosion.

Hoseman William Gorman, age 32: died February 14, 1926, when he became trapped during a major fire at Ben's Bakery, Pepperell Street. He died of smoke inhalation. Deputy Chief [Joseph] Harber attempted a rescue but was unable to find him because of intense smoke.

Hoseman William Cormier, age 33: died April 24, 1930, when he fell from a ladder during a Barrington Street fire. Another firefighter, Nelson Cormier, was on the ladder as well and both were thrown to the ground when a passing car caught the hose line.

Hoseman William Knapman, age 62: died January 25, 1939, when he suffered a heart attack while carrying out firefighting operations at a Brunswick Street fire. He was taken into a nearby building but did not respond to treatment.

Hoseman William Boston, age 62: died May 1, 1954, when he suffered a heart attack while working a grass fire in Halifax County. He was a member of Fairview Fire Department that became part of Halifax during amalgamation in 1969.

Captain Earl Fox, age 38: died January 15, 1956, when he suffered a heart attack during a major fire on Green Street. He was Station Captain at the Morris Street fire station No. 3.

Captain Richard Keily, age 50: died December 12, 1960, when he suffered a heart attack while directing his crew during a 2nd alarm fire at the Halifax Infirmary, Queen Street. He was rushed inside the hospital for treatment but did not respond. He was Rescue Car Captain at the Morris Street fire station No. 3.

Lieutenant William Carter, age 43: died December 3, 1973, after suffering a heart attack while returning from a call. He was the Officer on Rescue 2 and collapsed in the front seat. The crew began resuscitation immediately with no response.

Firefighter Allen MacFarlane, age 51: died January 20, 1980, when he suffered a heart attack during a fire on Gottingen Street. He was inside the building, suffered smoke inhalation and collapsed outside. He was Platoon Chief's Aide at West Street Station No. 2.

Epilogue:

Fireman's Prayer and History

Our faith comes in moments ... yet there is a depth in those brief moments which constrains us to ascribe more reality to them than to all the other experiences.

– Ralph Waldo Emerson

When my father passed in 2006, I was honoured by the large number of former and current Halifax firefighters who showed up to pay their final respects. I had contacted Tom Silver, who had previously worked as a media spokesperson for the fire department and was an old friend of mine, to inform him of my father's death. Silver arranged to have an electronic bulletin sent out to all metro fire stations announcing Landry's passing.

My father's wake was held, for one day, at Cruickshank's Funeral Home, which at the time was located on Robie Street, in an area dubbed "The Willow Tree." The intersection of multiple lanes and streets is one of the busiest in the city and is always congested during peak traffic times.

I had arranged two visitations that day, one in the afternoon and one in the evening. During the daytime, from two until four, a large contingent of current firefighters turned out, as well as many of my father's acquaintances and retired firefighting friends.

At one point, I was chatting with a group of the younger first responders when one of them, whom I did not know and who did not personally know my father, pulled me aside and asked me to come outside on the front steps of the funeral home to see something.

There, lining Robie Street, parked directly outside, were five or six fire trucks which had been driven by firefighters from stations across Halifax to the funeral home. It was an emotional sight to see so many of them parked together in a small area when there was no emergency happening. They signalled to passersby that one of their own lay inside.

It may sound strange, and it did to one of my family members, who reacted oddly when I told him that, despite my intense grief, at that exact moment, I was happy; I was very moved and pleased for my father that so many firefighters cared enough to show up, and did so with such flair and dignity.

The show of support grew from there.

The Halifax Regional Fire & Emergency Honour Guard, organized by local 268 of the IAFF, volunteered to serve at Landry's funeral. Union president Paul Boyle and secretary Brad Connors organized the honour guard.

They stood silently, carrying flags, at the entrance of our parish, St. Agnes Church, as we walked up the stone steps and while my father's casket was carried inside. During part of the service, the guards stood at attention at the front of the church, near the altar. Father Lloyd O'Neil, a former fire and police chaplain, along with our own priest, Father John Williams, who has also served as a pastor to the Canadian Armed Forces, co-officiated during my father's funeral mass. They are both friends of my family. It was touching and special to have them both agree to officiate. Father Williams gave a moving homily about the selflessness of all first responders which moved many, including me, to tears. The church was packed with several hundred people.

My father would not have expected any of it.

My husband, Rob Dauphinee, whom Landry regarded as the son he never had, spoke passionately and with conviction to the large congregation about his father-in-law. This is Dauphinee's verbatim funeral speech, which I have saved:

On behalf of the family, I'd first like to thank you all for coming here today in the presence of God to celebrate the life of Baz Landry. He would be so pleased to see you all here today.

My father-in-law was the kind of man who gave of himself all the time. He gave a lot of laughs, to his card buddies at the Brightwood Golf and Country Club in Dartmouth. He gave a good golf game, to his buddies at Ashburn and Brightwood. He gave tirelessly to his family, especially to his wife Theresa, loving daughter, Janice, to whom he taught the value of the work ethic and believing in yourself, and his six-year-old, only granddaughter Laura, on whom the sun rose and set.

And finally he gave it all, risking his life, time and time again, fighting fires in Metro Halifax, for thirty years. Baz was a good firefighter. One of the boys you might say. He loved to tell stories about the good old days at the firehouse and some of the practical jokes the firemen would play on each other. He also told stories, although these ones not as often, about the times he almost didn't make it out of the flames and smoke.

On one occasion, he climbed a trellis on a house in Halifax and swung up onto a window ledge. He climbed into the burning house, without any breathing apparatus, crawled across the floor of a bedroom, feeling his way, since it was full of smoke, to find a small baby boy in his crib. He carried the boy to the window and gave him mouth-to-mouth, saving the small child's life, but in this selfless act, almost losing his.

For this bravery, he was named Canadian Firefighter of the Year in 1979 and a year later, in … 1980, he was decorated at Rideau Hall in Ottawa with a Medal of Bravery for that heroic act. Over the years, Baz never talked about this honour with many people. He was a very humble and gracious man.

When he got sick, within the last year, and his lungs eventually became affected, he once said that maybe running into all those fires without breathing apparatus (because regulations weren't as stiff in those days) was finally taking its toll.

For that reason and many others, we, as his family and friends who put their faith in The Lord, know that he is finally at rest, in a place without smoke and fire, where people who put their lives on the line are recognized as being the true heroes they are, and, where today, at this very hour, we hope that Baz Landry is out on the 17th tee, swinging away, without any pain at all.

Dauphinee delivered his speech with the perfect balance of emotion and dignity. It was a final and fitting tribute and farewell to Landry by someone who really knew and loved him. Once he finished and descended from the altar, I immediately rose to my feet and hugged my husband, openly crying. Many others broke down with us.

I have also kept the guest registry that I placed at both the church and the funeral home; its signatures act as a permanent record of the many people who attended. Upon rereading it, long after I started my research, I discovered the names of many current and retired Halifax firefighters and, significantly, some of the same first responders mentioned in this book or whom I have since interviewed. Six years before I started working on this project, they were also present at Landry's wake and funeral. Those signatures are from Bernie Harvey, Dave Singer, Frank Baker, Don Snider and the late Gordon "Champ" McIsaac, who held court at Dad's wake, telling us wonderful stories.

While the 2006 funeral mass was taking place, we were unaware that the honour guard had also decided to continue on to Landry's burial site. They were waiting, at their post, as we arrived in a lengthy funeral procession. I will never forget the sight of two fire trucks parked in the graveyard. The sentries stood silently by as we said our difficult goodbyes. It was a send-off for a man who would have been humbled by the show of appreciation, love and support.

Through streams of tears, I approached the flag bearers and thanked them for accompanying my father to his final resting place. They did not speak, as they are not allowed to while officiating. I know they heard me because I looked right at them when I spoke. I saw the emotion on their faces.

During that initial time of mourning, one of my father's firefighting friends, who had attended at the funeral home, quietly approached me and handed me a prayer card with the famous "Fireman's Prayer" embossed on it. I still have the card he gave me. As a sign of respect to him, my father's friends, colleagues and supporters and all first responders, I have chosen to share this version of it near the closing of this book.

The prayer and world-renowned poem is written by the late Alvin William (Smokey) Linn, who was a firefighter with the Wichita Fire Department in Kansas. He wrote it after responding to a horrific fire in 1958 involving children trapped in a burning apartment building.

"The Fireman's Prayer" was originally published in a book called *A Celebration of Poets* the year in which the fire occurred. One online article explaining its history quotes a speech given by Linn's granddaughter, Penny McGlachlin. She explains the lasting impact that fire call had on her famous, fallen grandfather:

"… The firefighters could see the children in the windows but could not rescue them due to the iron bars that the apartment owner had installed. All they could do was try to contain the fire. About one in the morning, Smokey found himself sitting at the station's kitchen table putting in to words the emotions inside of him from that evening. The following words are one man's prayer … He was a husband, father, grandfather, and a son who knew how precious and short life can be."

At the time, Linn reportedly had children around the same age as the victims. McGlachlin is also quoted as saying there were no grief counsellors to support the first responders who attended at the apartment fire. This prayer and poem was part of Linn's way of coping and expressing himself during its aftermath.

Linn died March 31, 2004.

The Fireman's Prayer

A.W. "Smokey" Linn

When I am called to duty, God
Whenever flames may rage,
Give me the strength to save some life
Whatever be its age.
Help me to embrace a little child
Before it's too late,
Or some older person
From the horror of that fate.
Enable me to be alert
And hear the weakest shout,
And quickly and efficiently
To put the fire out.
I want to fill my calling
And give the best in me,
To guard my neighbor
And protect his property.
And if according to Your will
I have to lose my life,
Please bless with Your protecting hand
My children and my wife.

Atlantic Canada Firefighters' Memorial Service

Late into the work for this book, I had the distinct honour of speaking at the second annual Atlantic Canada Firefighters' Memorial Service held at St. Agnes Church on June 2, 2013. It was the first time I came face to face with members of the HRFE's Honour Guard since my father's 2006 funeral.

I was pleased the white-gloved guard members sat in the church pews located directly in front of me. That meant when it came time for me to deliver the following speech, from one of the altar lecterns, I could look right at them.

It gave me the chance I had been waiting for – to personally thank them for what they had done for the late Baz Landry, and for what they continue to do for many other families of first responders.

I said:

It's a great honour for me to be speaking at today's Firefighters' Memorial Service – on behalf of loved ones and families of those who have passed, who are retired or who are currently serving.

My late father, Baz Landry, was a veteran Halifax firefighter. He retired in 1988 after thirty-one years of service. Some of you worked with Baz and knew him. He loved being a firefighter. My dad died seven years ago. This was his parish. Our whole family attends St. Agnes. It was here that my father's funeral was held, in 2006, with a full honour guard, like the one serving today. What a blessing they are to so many families. We thank them for being there, for all of us, when it counts.

Family is why we are here. We are all family, in one way or another. You, as firefighters and first responders, share a special bond with each other. That bond happens among people when they routinely risk everything to protect people, property, and the community. For that, we are deeply grateful.

You, working as firefighters and first responders, never know what you're going to face when the alarm sounds. When everyone else is running away, you run towards the emergency. Despite the uncertainty, you do not hesitate. For that, we are deeply grateful.

You do not get the recognition you deserve and many people do not understand or appreciate the toll this work can take: both mentally and physically. Yet you come back – day after day. For that, we are deeply grateful.

I am a journalist, and over the past eighteen months have been researching and writing a magazine series and book about the work you do. Both projects are dedicated to all firefighters, first responders, and my late father, who was a Canadian Medal of Bravery recipient. Some of you here today have spoken with me or have helped me. On behalf of my family, thank you.

The purpose of my project is to underline the selflessness and dedication of our fallen, retired, and current firefighters and first responders – people like you who deserve to be recognized and thanked.

I am leaving you with a quote, a favourite of mine that is in the book. It's about success.

We must re-evaluate how we define it.

And it reads:

> *To laugh often and much*
> *To win the respect of intelligent people and the affection of children*
> *To earn the appreciation of honest critics and endure the betrayal of false friends*
> *To appreciate beauty*
> *To find the best in others*
> *To leave the world a bit better, whether by a healthy child, a garden patch or a redeemed social condition*
> *To know even one life has breathed easier because you have lived*
> *This is to have succeeded.*

To our fallen firefighters and first responders, our retired veterans, and those of you currently serving: you are the definition of what it is to be successful. For that – we are deeply grateful.

FFAP Contact Information

If you or a loved one would like a referral, information about the Halifax Regional Fire & Emergency's Firefighters & Family Assistance Program (FFAP) or someone to talk to who understands, please contact:

Paul MacKenzie, co-ordinator
902-490-6271 (office) 902-483-6068 (cell)
mackenp@halifax.ca (e-mail)
Confidentiality is assured.

Acknowledgements

I would like to express my profound respect for and gratitude to the entire Gibson/McKenzie family, FFAP coordinator and former police officer Paul MacKenzie, as well as the many Halifax fire-fighters interviewed for this work. All of their names appear in the acknowledgements.

Every single person has been kind and welcoming during our conversations, whether in person, via telephone, e-mail or social media. They have frankly discussed experiences that are of a deeply personal nature. I have tried my best to represent them delicately yet factually.

Twenty-one people have been interviewed for *The Sixty Second Story*; their willingness to talk is testament to their concern for others and a willingness to try to help make a difference. I hope these stories shed more light on the exact nature of what it is truly like to be a first responder.

There are a number of people I would like to publicly thank for their support, encouragement and assistance in terms of getting this book done.

The order of names does not signify importance; they all helped me in some way, and, therefore, I am grateful to the following people, starting with my immediate family:

Rob Dauphinee, Laura Dauphinee, Theresa Landry, Betty Dauphinee, Halifax Regional Fire & Emergency fire chief Doug Trussler, Darlene Ellis, Linda Dodge, Don Snider, IAFF Local 268, Chris Camp, Firefighters Family Assistance Program, Paul MacKenzie, Yvonne Colbert, Rick Howe, Amber Leblanc, Peter Mallette, Margaret McGee, Bill Jessome, *Feedline Magazine*, John Giggey, Theresa Rath, *Halifax Magazine*, Trevor J. Adams, Afton Doubleday, Joe Murphy, Bruno Lemay, Nancy Fey, Marie-Pierre Bélanger, Tanya Thomson, Bernie Harvey, Gerry Condon, Rob Brown, Don Swan, Ron Horrocks, Dave Singer, Doug Findlay, Bob Whorrall, John Fitzgerald, Joe Young, Frank Zwicker, Les Power, Heather Brown Harroun, Beth Mader, Tom Silver, Paul Boyle, Brad Connors, Father Lloyd O'Neil, Father John Williams, the Halifax Regional Fire & Emergency Honour Guard, Charlene McKenzie Meade, Nick McKenzie, Kim Gibson, Shirley McKenzie, Leanne McKenzie, the residents of 3390 Federal Avenue, Judy and David "Pops" Mamye, Paul Darrow, Julia Swan, Peggy Amirault, and Lesley Choyce.

Finally, in closing, I would like to thank my beloved father, Basil "Baz" Landry, for entrusting his Medal of Bravery and collection of historical documents and media reports to me.

I did know what to do with them, Dad.

Bibliography

Interviews (in the order they were done):

John Fitzgerald. Personal interview, January 16, 2012.

Don Swan. Personal interview. April 25, 2012.

Don Snider. Personal interviews. May 1, 2012 and May 23, 2012.

Les Power. Telephone interview. May 3, 2012.

Frank Zwicker. Telephone interview. May 3, 2012.

Ronald Horrocks. Telephone interview. May 4, 2012.

Dave Singer. Telephone interview. May 6, 2012.

Paul MacKenzie. Personal interview. May 7, 2012.

Joe Young. Personal interview. May 14, 2012.

Kim Gibson. Personal interview. May 23, 2012.

Doug Findlay. Phone interview. June 4, 2012.

Gerry Condon. Personal interview. June 11, 2012.

Bernie Harvey. Personal interview. June 13, 2012.

Chris Camp. Personal interview. June 14, 2012.

Bob Whorrall. Telephone interview. December 6, 2012.

Charlene McKenzie Meade. Telephone interview. December 7, 2012.

Shirley McKenzie. Personal interview. December 14, 2012.

Nick McKenzie. Personal interview. December 14, 2012.

Heather Brown Harroun. Telephone/E-mail/Facebook interviews. February 11-12, 2013.

Beth Mader. E-mail interview. February 12, 2013.

Rob Brown. Personal interview. April 30, 2013.

Articles from a daily newspaper (alphabetically):

Author Unknown. Story title unknown. Billy Wells interview excerpts. *The Mail Star*. (Halifax). December 6, 1967. Section and page unknown. Print.

Author Unknown. "Firefighter commended for bravery." *The Mail Star*. (Halifax). October 16, 1978. News section, front page. Print.

Bernard, Elissa. "Halifax Fireman is honoured for rescue." *The Mail Star*. (Halifax). Day and Month unknown 1979. Section and page unknown. Print.

DeMings, John. "Coast Guard crew earns medals." *The Digby Courier*. (Digby). February 8, 2013. Section and page unknown. Print.

MacDonald, Marilyn. "Fireman wins Medal of Bravery." *The Mail Star*. (Halifax). Day and Month unknown 1980. Section and page unknown. Print.

Miller, Peggy. "Firefighter wins another award." *The Mail Star*. (Halifax). Day and month unknown 1979. Section and page unknown. Print.

Schneidereit, Paul. "Routine fire call ended in deadly Halifax Explosion." *The Chronicle Herald*. (Halifax). Date, section and page unknown. Print.

The Canadian Press. "Double murder, suicide near Ottawa involved two children." *The Chronicle Herald*. (Halifax). January 13, 2013. Section and page unknown. Print.

Articles from additional publications (by their year):

Jones, Robert L., *Canadian Disasters – An Historical Survey.* Date unknown. Print.

Author unknown. *Decorations for Bravery*. Honours Secretariat, Government House. Date unknown. Print.

Author unknown. *A Guide to the Wearing of Orders, Decorations and Medals.* The Chancellery of Canadian Orders and Decorations, Government House. Date unknown. Print.

Snider, Don and Palm Communications. *An Historical Celebration – 225 Years of Firefighting in Halifax (1768-1993)*. Date unknown. Print.

Author unknown. Dispatcher's Report: Response Number 782823. October 2, 1978. Print.

McIsaac, Gordon. Fire Report. October 2, 1978. Print.

McIsaac, Gordon. Details of Firefighting Operations. October 2, 1978. Print.

Whorrall, Bob. Details of Firefighting Operations. October 2, 1978. Print.

Author unknown. Details of Fire Fighting Operations. Halifax Fire Department. October 2, 1978. Print.

Landry, Basil. Details of Firefighting Operations. October 2, 1978. Print.

Findlay, Doug. Chief Officer's Report – 2, Summary of Operations and Story of Fire. October 3, 1978. Print.

Findlay, Doug. Chief Duty Officer Report. October 3, 1978.

Starrett, Dave. Fire Prevention Officer's Report. October 5, 1978. Print.

Horrocks, Ron. Nomination form: Fireman's Fund Insurance Bureau of Canada. Day unknown. October 1978. Print.

Author unknown. Media Release. Fireman's Fund Insurance Company of Canada. August 22, 1979. Print.

Horrocks, Ron. Nomination Form: International Association of Fire Chiefs. Day unknown. August 1979. Print.

Horrocks, Ron. City Council Briefing Notes. July 11, 1979. Print.

Horrocks, Ron and staff. Halifax Fire Department Annual Report 1979, Office of the Fire Chief. Day unknown. April 1980. Print.

Crosby, Howard. Constituency Report. Winter 1980-1981. Print.

McIsaac, Gordon, Daily Report. February 19, 1980. Print.

Author unknown. *Investiture.* Government of Canada. September 26, 1980. Print.

McIsaac, Gordon. Official Fire Report. Day unknown. February 1980. Print.

Author unknown. *Recipients of Decorations for Bravery.* Government of Canada. Day unknown. January 1981. Print.

Singer, Dave. "The Halifax Explosion – Day of Destruction." *The Seniors Advocate.* Day and month unknown. 1987. Print.

Hallessey, Cathy. "CEPU Local Union Officer Wins Award for Bravery at Westray." Day unknown. February 1995. Print.

Dauphinee, Rob. Speech. May 2006. Print.

Musiack, Susan. "Medal of Bravery from Westray on e-bay auction block." September 29, 2010. Print.

Bélanger, Marie-Pierre. Media Release. Governor General to Present 50 Decorations for Bravery. February 7, 2013.

Landry, Janice. Speech. June 2, 2013. Print.

Information from websites (by dates retrieved):

Author unknown. *Fireman's Prayer.* Retrieved from www.aspiringfirefighters.com April 30, 2012.

Author unknown. *Fireman's Prayer.* Retrieved from www.namesandthingszone.com April 30, 2012.

Author unknown. *MS Caribe.* www.scotiaprince.com and www.cruisecritic.com. Retrieved May 7, 2012.

Author unknown. *LaSalle Heights Disaster.* Retrieved from Wikipedia. org. January 15, 2013.

Author unknown. www.parl.gc.ca. Retrieved information on Howard Crosby. April 30, 2012.

Author unknown. www.gg.ca. Retrieved information on The Medal of Bravery and Governors General. Date not recorded.

Author unknown. www.osha.gov/SLTC/emergencypreparedness/ guides/critical.html. Retrieved information on critical stress. Date not recorded.

Author unknown. www.stagnesparishrc.ca. Retrieved information about Father John Williams. January 28, 2013.

Author unknown. *The Fireman's Prayer.* www.southlinefire.com. Retrieved February 11, 2013.

McGlachlin, Penny. Speech excerpts. www.fellowshipofchristianfirefighters.com. Retrieved February 11, 2013.

Ralph Waldo Emerson. *Complete Works of RWE.* www.rwe.org.
Retrieved February 11, 2013.

Ralph Waldo Emerson. *Quotes by Emerson, Ralph Waldo.*
www.googlebooks.com. Retrieved February 11, 2013.

Personal Letters (alphabetically):

Author Unknown. Letter to Ronald M. Horrocks. October 5, 1979.
Transcript.

Edmund Morris. Letter to Basil Landry. July 12, 1979. Handwritten.

J.D. Beck. Letter to Basil Landry. November 4, 1979. Transcript.

Joe Clark. Letter to Basil Landry. September 22, 1980. Transcript.

Howard Crosby. Letter to Basil Landry. September 29, 1980.
Transcript.

Roger de C. Nantel. Letter to Basil Landry. July 21, 1980. Transcript.

Articles from a magazine (alphabetically):

Atkinson. Walter. "Halifax Fire Fighter (sic) is Honoured for
Rescue." *International Firefighter-Canadian Edition.* No Volume
and Issue Numbers. (January/February 1980*)*: Pages 8-9.
Print.

Author unknown. "Halifax firefighter wins bravery award."
Firefighting in Canada. No Volume and Issue Numbers.
(October-November 1979): Page 29. Print.

Author unknown. "Halifax Fire Officer Wins Fireman's Fund
Award for Outstanding Bravery." *The Canadian Firefighter.* 3.16
(September-October 1979): Page 5. Print.

Author unknown. "Dec. 6, 1917," *Atlantic Firefighter.* Volume and
issue unknown. (November 1992): Pages 2-3. Print.

Smith, Dennis. Article unnamed. *Firehouse Magazine.* Volume and
   Issue unknown. (November 1979): Pages unknown. Print.

E-mails (alphabetically):

Bélanger, Marie-Pierre. "Medal of Bravery. Basil Landry 1980."
   Message to the author. April 25, 2012. E-mail.

Bélanger. Marie-Pierre. "Fact check – NS Medals of Bravery from
   last week." Message to author. February 11, 2013.

Camp, Chris. "Thursday 11am." Message to the author. June 8,
   2012. E-mail.

Camp, Chris. "Update." Message to the author. September 17, 2012.
   E-mail.

Colbert, Yvonne. "Amber." Message to the author. January 17, 2013.
   E-mail.

Dodge, Linda. "Interview with Chief." Message to the author. June
   7, 2012. E-mail.

Dodge, Linda. "MB recipients from Halifax Fire." Message to the
   author. September 29, 2012. E-mail.

Dodge, Linda. "Fact checking." Message to the author. December 12,
   2012. E-mail.

Doubleday, Afton. "Help requested." Message to the author. April
   30, 2012. E-mail.

Doubleday, Afton. "MB fact check request." Message to the author.
   September 25, 2012. E-mail.

Doubleday, Afton. "Westray Fact Check." Message to the author.
   December 17, 2012. E-mail.

Doubleday, Afton. "Book-last fact checks." Message to author.
   February 11, 2013. E-mail.

Ellis, Darlene. "Request of report from 1978. Basil Landry, MB recipient." Message to the author. May 2, 2012. E-mail.

Fitzgerald, John. "Couple Things." Message to the author. December 3, 2011. E-mail.

Fitzgerald, John. "Meeting next week." Message to the author. January 10, 2012. E-mail.

Fitzgerald, John. "Next week. Monday?" Message to the author. January 14, 2012. E-mail.

Fitzgerald, John. "Underway." Message to the author. April 23, 2012. E-mail.

Harroun, Heather. "Thank you and questions." Message to the author. February 11, 2013. E-mail.

MacKenzie, Paul. "Monday confirmed." Message to the author. May 2, 2012. E-mail.

MacKenzie, Paul. "Request." Message to the author. May 2, 2012. E-mail.

MacKenzie, Paul. "Fact Check." Message to the author. September 17, 2012. E-mail.

MacKenzie, Paul. "Fact Checks." Message to the author. January 16, 2013. E-mail.

Mader, Beth. "Thank you and questions." Message to the author. February 12, 2013. E-mail.

Rath, Theresa. "Former Chief MacDonald." Message to the author. January 16, 2013. E-mail.

Snider, Don. "Meeting." Message to the author. May 27, 2012. E-mail.

Snider, Don. "Drop off and update." Message to the author. September 19, 2012. E-mail.

Snider, Don. "MB recipients from Halifax Fire." Message to the author. September 24, 2012. E-mail.

Snider, Don. "Fact check for first issue." Message to the author. October 10, 2012. E-mail.

Snider, Don. "Fact checking for book on boy's rescue." Message to the author. November 23, 2012. E-mail.

Snider, Don. "Fact checking for book on boy's rescue." Message to the author. November 24, 2012. E-mail.

Thomson. Tanya. "Couple Things." Message to the author. December 2, 2011. E-mail.

Thomson, Tanya. "Dad." Message to the author. March 7, 2012. E-mail.